Multiplication table games

MATH GAMES FOR KIDS

by: Laura Putman, Bright Minds Engaged

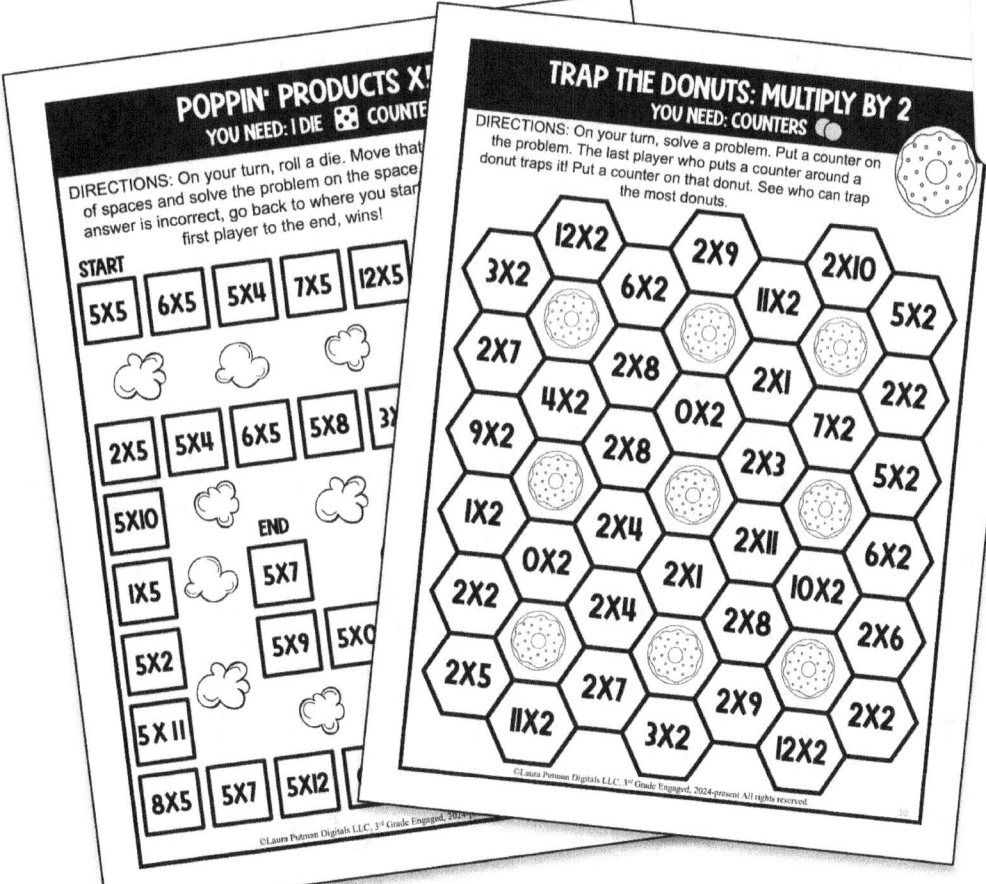

© Laura Putman, Bright Minds Engaged, 2024-present, All rights reserved.

All images created by Laura Putman Digitals LLC. All rights reserved. No part of this publication may be reproduced, distributed, or transmitted in any form or by any means. This includes photocopying, recording, or other electronic or mechanical methods without prior permission of the publisher, except in the case of brief quotations embodied in critical reviews and other noncommercial uses permitted by copyright law.

Clipart by Educlips – www.educlips.com
Partyhead Graphics
Johnny's Clips
Hidesy's Clipart

No part of this product maybe used or reproduced for commercial use.

Contact the author :
laura@thirdgradeengaged.com

MULTIPLICATION FACTS

Parent Strategy Guide

Get the most out of this workbook!

✓ Tips for Success
✓ Strategy Posters
✓ Progress Chart

Scan here to get it!

The guide to helping your child master multiplication facts

By Laura Putman, M.Ed

© Laura Putman, Bright Minds Engaged, 2026-present, All rights reserved.

TABLE OF CONTENTS

p. 4-8 – multiplying by 0 and 1

p. 9-17 – multiplying by 2

p. 18-26 – multiplying by 3

p. 27-35 – multiplying by 4

p. 36-44 – multiplying by 5

p. 45-53 – multiplying by 6

p. 54-62 – multiplying by 7

p. 63-71 – multiplying by 8

p. 72-80 – multiplying by 9

p. 81-87 – multiplying by 10

p. 88-95 – multiplying by 11

p. 96-103 – multiplying by 12

p. 104-111 – multiplying mixed facts

ROLL & SOLVE: MULTIPLY BY 0 & 1

YOU NEED: 1 DIE 🎲 CRAYONS 🖍️

DIRECTIONS: Assign one player even numbers on the die and the other player odd numbers. Take turns rolling. If a player rolls one of their numbers on the die, they solve the next problem under that die and color the space. If they do not roll one of their numbers, their turn is over. See who can fill their columns first!

⚀	⚁	⚂	⚃	⚄	⚅
1X1	0X5	7X1	0X10	8X0	2X1
0X12	1X11	10X1	1X8	1X3	0X0
5X1	1X8	0X4	1X3	12X1	7X1
0X2	0X1	9X1	4X0	0X7	0X9
4X1	3X0	5X1	11X0	0X3	1X6
1X0	12X1	6X0	1X3	1X6	11X0

© Laura Putman, Bright Minds Engaged, 2024-present, All rights reserved.

TRAP THE SPIDERS: MULTIPLY BY 0 & 1
YOU NEED: COUNTERS

DIRECTIONS: On your turn, solve a problem. Put a counter on the problem. The last player who puts a counter around a spider traps it! Put a counter on that spider. See who can trap the most spiders.

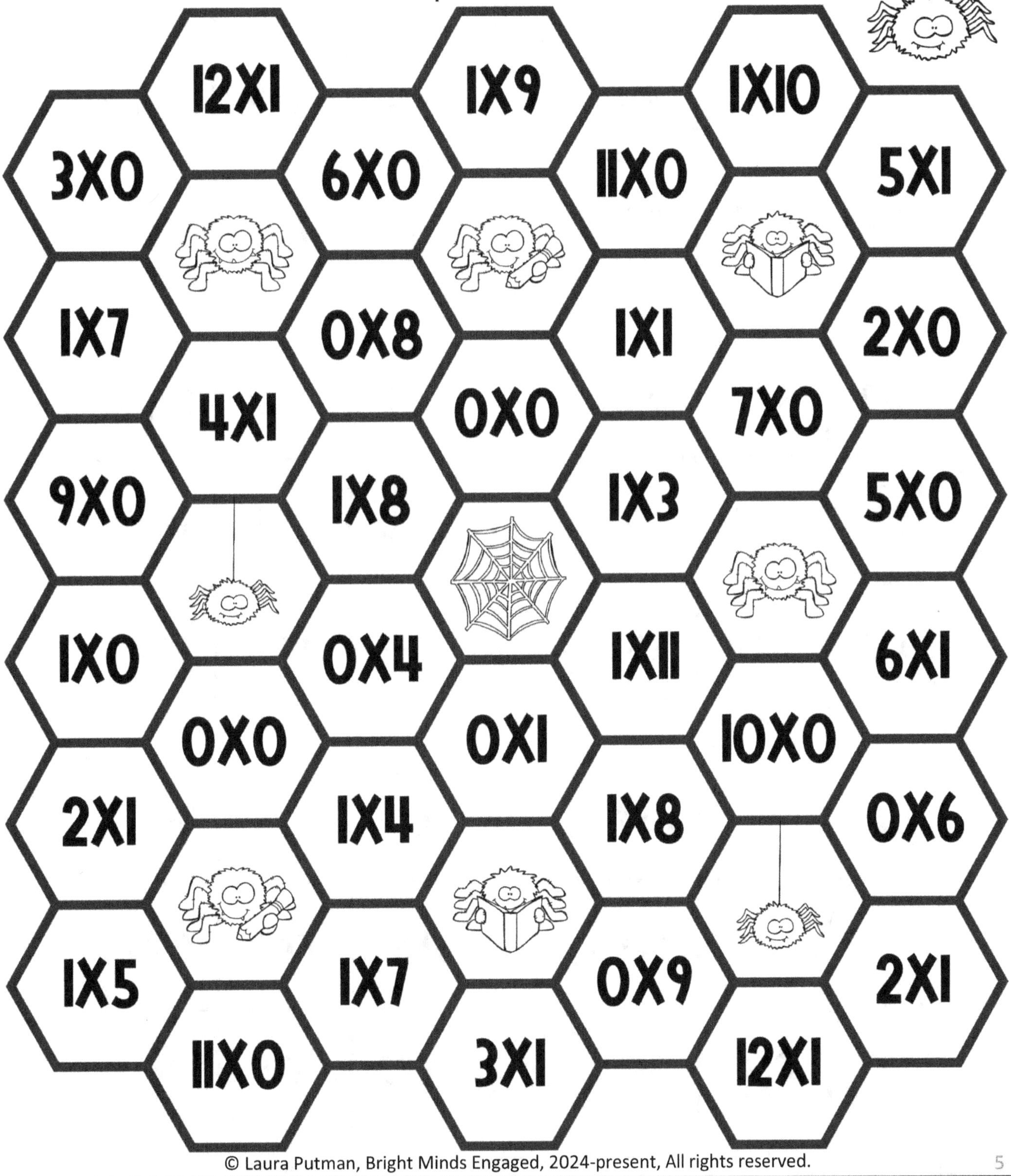

- 12X1
- 1X9
- 1X10
- 3X0
- 6X0
- 11X0
- 5X1
- 1X7
- 0X8
- 1X1
- 2X0
- 4X1
- 0X0
- 7X0
- 9X0
- 1X8
- 1X3
- 5X0
- 1X0
- 0X4
- 1X11
- 6X1
- 0X0
- 0X1
- 10X0
- 2X1
- 1X4
- 1X8
- 0X6
- 1X5
- 1X7
- 0X9
- 2X1
- 11X0
- 3X1
- 12X1

© Laura Putman, Bright Minds Engaged, 2024-present, All rights reserved.

TIC-TAC-TOE: MULTIPLY BY 0 & 1

DIRECTIONS: Play a game of tic-tac-toe! Before you mark a space as yours, you must solve the problem in that space.

0X0	12X1	1X1		6X0	7X1	1X2
0X7	0X9	11X1		0X1	8X0	3X1
1X10	0X8	1X5		9X1	1X0	0X4

1X9	5X0	1X10		4X1	1X3	12X0
0X1	1X8	2X0		5X1	1X12	0X11
1X3	0X7	0X12		7X1	1X0	8X0

0X7	0X1	0X5		1X6	0X3	4X1
1X6	10X0	1X4		11X1	10X1	0X0
2X1	0X11	3X1		0X2	6X1	9X1

© Laura Putman, Bright Minds Engaged, 2024-present, All rights reserved.

FOUR PROBLEMS IN A ROW: MULTIPLY BY 0 & 1

YOU NEED: CRAYONS OR COUNTERS

DIRECTIONS: On your turn, solve a problem and color it or cover it. Each player uses a different color. The first player to get 4 in a row wins!

0 X 2	0 X 1	9 X 0	6 X 1	0 X 1	1 X 7
5 X 1	10 X 0	12 X 0	4 X 1	0 X 8	1 X 2
7 X 0	1 X 8	0 X 0	1 X 3	1 X 4	1 X 0
6 X 1	0 X 11	1 X 7	6 X 0	1 X 1	11 X 1
0 X 2	5 X 1	9 X 0	1 X 8	10 X 0	3 X 1
1 X 12	11 X 0	6 X 1	9 X 0	1 X 5	4 X 0

© Laura Putman, Bright Minds Engaged, 2024-present, All rights reserved.

MATH BATTLE: X0 & 1

YOU NEED: 1 DIE **COUNTERS**

DIRECTIONS: On your turn, roll a die. Move that number of spaces and solve the problem on the space. If your answer is incorrect, go back to where you started. The first player to the end, wins!

START

0X3		0X0	8X1	0X7		0X9	1X6
7X1		9X1		10X1		3X1	
0X11		0X5		0X11		1X10	
4X1		1X2		2X1		4X0	
0X6		0X8		1X0		1X1	
10X1		12X1		1X0		7X0	
0X12		0X9		1X6		1X11	
1X1	5X0	4X1		8X1	3X0	1X2	

END (top right, after 1X6)

ROLL & SOLVE: MULTIPLY BY 2
YOU NEED: 1 DIE CRAYONS

DIRECTIONS: Assign one player even numbers on the die and the other player odd numbers. Take turns rolling. If a player rolls one of their numbers on the die, they solve the next problem under that die and color the space. If they do not roll one of their numbers, their turn is over. See who can fill their columns first!

⚀	⚁	⚂	⚃	⚄	⚅
2X1	2X5	7X2	2X0	8X2	2X3
2X12	2X11	10X2	2X8	2X9	0X2
5X2	2X8	2X4	3X2	12X2	7X2
1X2	1X2	2X0	4X2	2X7	2X9
4X2	9X2	5X2	2X11	2X2	2X6
2X10	12X2	6X2	2X3	2X6	11X2

©Laura Putman Digitals LLC, 3rd Grade Engaged, 2024-present All rights reserved.

TRAP THE DONUTS: MULTIPLY BY 2
YOU NEED: COUNTERS

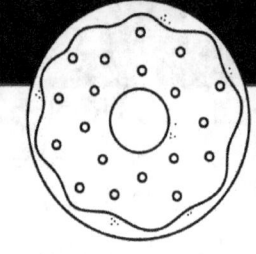

DIRECTIONS: On your turn, solve a problem. Put a counter on the problem. The last player who puts a counter around a donut traps it! Put a counter on that donut. See who can trap the most donuts.

	12X2		2X9		2X10
3X2		6X2		11X2		5X2
2X7		2X8		2X1		2X2
	4X2		0X2		7X2	
9X2		2X8		2X3		5X2
1X2		2X4		2X11		6X2
	0X2		2X1		10X2	
2X2		2X4		2X8		2X6
2X5		2X7		2X9		2X2
	11X2		3X2		12X2	

TIC-TAC-TOE: MULTIPLY BY 2

DIRECTIONS: Play a game of tic-tac-toe! Before you mark a space as yours, you must solve the problem in that space.

0X2	12X2	2X1
2X7	2X9	11X2
2X10	2X8	2X5

6X2	7X2	2X2
0X2	8X2	3X2
9X2	1X2	2X4

2X9	5X2	2X10
2X1	2X8	2X2
2X3	2X7	2X12

4X2	2X3	12X2
5X2	2X12	2X11
7X2	1X2	8X2

2X7	0X2	2X5
2X6	10X2	2X4
2X2	2X11	3X2

2X6	2X3	4X2
11X2	10X2	2X0
2X2	6X2	9X2

SPIN A PROBLEM: MULTIPLY BY 2
YOU NEED: PAPERCLIP / PENCIL / CRAYONS

DIRECTIONS: Use a paperclip and pencil to make a spinner. On your turn, spin the paperclip. Multiply that number by 2 and color the product in the table. Each player uses a different color. If the product is not open, your turn is over. See who can solve the most problems!

2	14	24	8	4	16
8	18	12	10	20	6
22	4	10	16	2	8
12	24	10	20	18	22
24	6	18	4	14	12

© Laura Putman, Bright Minds Engaged, 2024-present, All rights reserved.

PIG IN A PEN: MULTIPLY BY 2
YOU NEED: PAPERCLIP PENCIL

DIRECTIONS: On their turn, each player spins a number. Multiply the number by 2 and say the answer. If the answer is correct, draw a line to connect 2 dots. When a player completes a box, they write their initial in the box. At the end of the game, boxes are worth 1 point, and boxes with a pig in them are worth 5 points!

ARRAYS? HOORAY! MULTIPLY BY 2
YOU NEED: CRAYONS 2 DICE

DIRECTIONS: On their turn, each player rolls both dice. Add the dice together and multiply by 2. Draw an array for the problem. Each player uses a different color to make their arrays. Write the multiplication problem inside the array. If your array won't fit, your turn is over. When no more arrays can be made, the game is over. Whoever makes the most arrays, wins!

MULTIPLES OF 2 HUNT
YOU NEED: CRAYONS

Help the skiers make it down the mountain. Count by 2's to follow a path. If you get to 24, start at 2 again. Color or dab the spaces until you get to the bottom.

FOUR PROBLEMS IN A ROW: MULTIPLY BY 2
YOU NEED: CRAYONS OR COUNTERS

DIRECTIONS: On your turn, solve a problem and color it or cover it. Each player uses a different color. The first player to get 4 in a row wins!

2 X 2	0 X 2	9 X 2	6 X 2	2 X 1	2 X 7
5 X 2	10 X 2	12 X 2	4 X 2	2 X 8	2 X 2
7 X 2	2 X 8	2 X 0	2 X 3	2 X 4	1 X 2
6 X 2	2 X 11	2 X 7	6 X 2	1 X 2	11 X 2
2 X 2	5 X 2	9 X 2	2 X 8	10 X 2	3 X 2
2 X 12	11 X 2	6 X 2	9 X 2	2 X 5	4 X 2

© Laura Putman, Bright Minds Engaged, 2024-present, All rights reserved.

MULTIPLICATION MYSTERY X2'S

YOU NEED: 1 DIE 🎲 COUNTERS ⬤⬤

DIRECTIONS: On your turn, roll a die. Move that number of spaces and solve the problem on the space. If your answer is incorrect, go back to where you started. The first player to the end, wins!

START

2X3		2X0	8X2	2X7		2X9	2X6
7X2		9X2		10X2		3X2	**END**
2X11		2X5		2X11		2X10	
4X2		2X2		2X2		4X2	
2X6		2X8		2X0		2X1	
10X2		12X2		1X2		7X2	
2X12		2X9		2X6		2X11	
1X2	5X2	4X2		8X2	3X2	2X2	

© Laura Putman, Bright Minds Engaged, 2024-present, All rights reserved.

ROLL & SOLVE: MULTIPLY BY 3

YOU NEED: 1 DIE CRAYONS

DIRECTIONS: Assign one player even numbers on the die and the other player odd numbers. Take turns rolling. If a player rolls one of their numbers on the die, they solve the next problem under that die and color the space. If they do not roll one of their numbers, their turn is over. See who can fill their columns first!

⚀	⚁	⚂	⚃	⚄	⚅
3X1	3X5	7X3	3X10	8X3	2X3
3X12	3X11	10X3	3X8	3X3	0X3
5X3	3X8	3X4	3X3	12X3	7X3
3X2	1X3	9X3	4X3	3X7	3X9
4X3	3X3	5X3	11X3	2X3	3X6
3X0	12X3	6X3	3X3	3X6	11X3

© Laura Putman, Bright Minds Engaged, 2024-present, All rights reserved.

TRAP THE YETIS: MULTIPLY BY 3
YOU NEED: COUNTERS

DIRECTIONS: On your turn, solve a problem. Put a counter on the problem. The last player who puts a counter around a yeti traps it! Put a counter on that yeti. See who can trap the most yetis.

	12X3		3X9		3X10	
3X3		6X3		11X3		5X3
3X7		3X8		3X1		2X3
	4X3		0X3		7X3	
9X3		3X8		3X3		5X3
1X3		3X4		3X11		6X3
	0X3		3X1		10X3	
3X2		3X4		3X8		3X6
3X5		3X7		3X9		3X2
	11X3		3X3		12X3	

TIC-TAC-TOE: MULTIPLY BY 3

DIRECTIONS: Play a game of tic-tac-toe! Before you mark a space as yours, you must solve the problem in that space.

0X3	12X3	3X1
3X7	3X9	11X3
3X10	3X8	3X5

6X3	7X3	2X3
0X3	8X3	3X3
9X3	1X3	3X4

3X9	5X3	3X10
3X1	3X8	3X2
3X3	3X7	3X12

4X3	3X3	12X3
5X3	3X12	3X11
7X3	1X3	8X3

3X7	0X3	3X5
3X6	10X3	3X4
3X2	3X11	3X3

3X6	3X3	4X3
11X3	10X3	3X0
2X3	6X3	9X3

SPIN A PROBLEM: MULTIPLY BY 3
YOU NEED: PAPERCLIP ✏ PENCIL ✏ CRAYONS

DIRECTIONS: Use a paperclip and pencil to make a spinner. On your turn, spin the paperclip. Multiply that number by 3 and color the product in the table. Each player uses a different color. If the product is not open, your turn is over. See who can solve the most problems!

3	21	36	12	6	24
12	27	18	15	30	9
33	6	3	24	3	12
18	36	15	30	27	33
30	9	27	6	21	18

PIG IN A PEN: MULTIPLY BY 3
YOU NEED: PAPERCLIP 📎 PENCIL ✏️

DIRECTIONS: On their turn, each player spins a number. Multiply the number by 3 and say the answer. If the answer is correct, draw a line to connect 2 dots. When a player completes a box, they write their initial in the box. At the end of the game, boxes are worth 1 point, and boxes with a pig in them are worth 5 points!

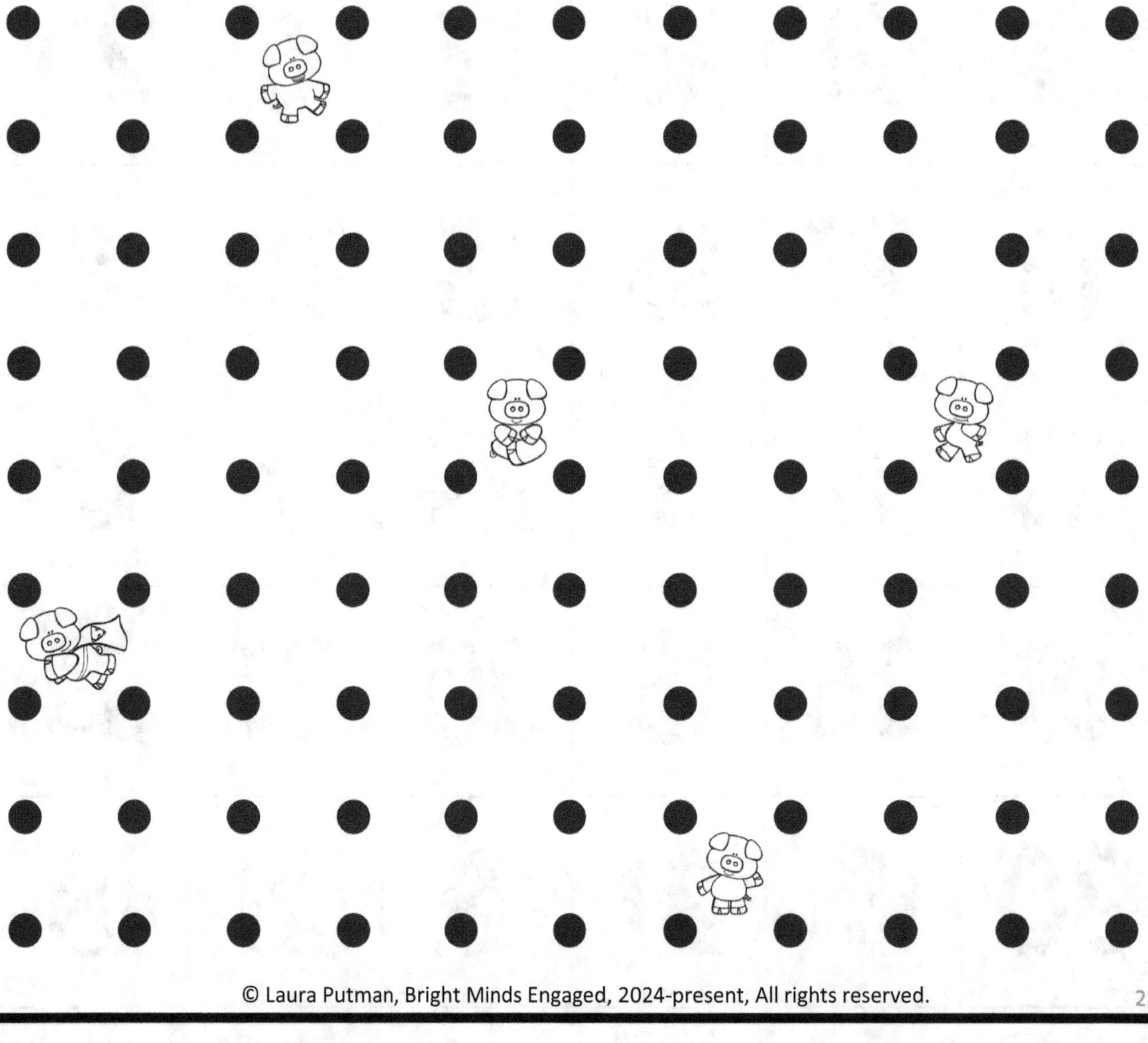

© Laura Putman, Bright Minds Engaged, 2024-present, All rights reserved.

ARRAYS? HOORAY! MULTIPLY BY 3
YOU NEED: CRAYONS 2 DICE

DIRECTIONS: On their turn, each player rolls both dice. Add the dice together and multiply by 3. Draw an array for the problem. Each player uses a different color to make their arrays. Write the multiplication problem inside the array. If your array won't fit, your turn is over. When no more arrays can be made, the game is over. Whoever makes the most arrays, wins!

MULTIPLES OF 3 HUNT
YOU NEED: CRAYONS

Help the bookworm get to the books. Count by 3's to follow a path. If you get to 36, start at 3 again. Color or dab the spaces until you get to the bottom.

START →

3	6	13	81	21	38	56	43
12	10	9	40	24	63	75	42
22	16	12	25	30	5	80	79
6	65	56	15	35	36	52	9
24	15	80	66	23	21	18	10	42	71	8	78
14	10	74	16	24	80	50	35	54	49	62	13
47	37	33	27	84	41	16	55	56	60	18	40
20	36	28	30	83	5	60	63	65	27	7	38
66	3	25	40	15	24	27	72
6	21	30	18	56	21	74	30
28	9	81	35	18	13	33	75
78	21	12	15	71	78	55	36

END →

FOUR PROBLEMS IN A ROW: MULTIPLY BY 3
YOU NEED: CRAYONS OR COUNTERS

DIRECTIONS: On your turn, solve a problem and color it or cover it. Each player uses a different color. The first player to get 4 in a row wins!

2 X 3	0 X 3	9 X 3	6 X 3	3 X 1	3 X 7
5 X 3	10 X 3	12 X 3	4 X 3	3 X 8	3 X 2
7 X 3	3 X 8	3 X 0	3 X 3	3 X 4	1 X 3
6 X 3	3 X 11	3 X 7	6 X 3	1 X 3	11 X 3
3 X 2	5 X 3	9 X 3	3 X 8	10 X 3	3 X 3
3 X 12	11 X 3	6 X 3	9 X 3	3 X 5	4 X 3

FOOTBALL FACTS X3'S

YOU NEED: 1 DIE  **COUNTERS**

DIRECTIONS: On your turn, roll a die. Move that number of spaces and solve the problem on the space. If your answer is incorrect, go back to where you started. The first player to the end, wins!

START

| 5X3 | 6X3 | 3X4 | 7X3 | 12X3 | 3X0 | 2X3 | 3X9 |

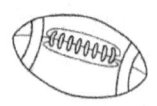   

| | | | | | | | 8X3 |

| 2X3 | 3X4 | 6X3 | 3X8 | 3X3 | 3X1 | | 3X3 |

 (22)

| 3X10 | | | | 3X10 | | | 10X3 |

 (44)

END

| 1X3 | | 3X7 | | | 4X3 | | 1X3 |

| 3X2 | | 3X9 | 3X0 | 11X3 | 5X3 | | 11X3 |

| 3X11 | | | | | | | 5X3 |

| 8X3 | 3X7 | 3X12 | 0X3 | 3X6 | 3X9 | 3X3 | 4X3 |

© Laura Putman, Bright Minds Engaged, 2024-present, All rights reserved.

ROLL & SOLVE: MULTIPLY BY 4
YOU NEED: 1 DIE • CRAYONS

DIRECTIONS: Assign one player even numbers on the die and the other player odd numbers. Take turns rolling. If a player rolls one of their numbers on the die, they solve the next problem under that die and color the space. If they do not roll one of their numbers, their turn is over. See who can fill their columns first!

⚀	⚁	⚂	⚃	⚄	⚅
4X1	4X5	7X4	2X4	8X4	4X3
4X12	4X11	10X4	4X8	4X9	0X4
5X4	4X8	4X4	3X4	12X4	7X4
1X4	1X4	4X0	4X4	4X7	4X9
4X4	9X4	5X4	4X11	4X2	4X6
4X10	12X4	6X4	4X3	4X6	11X4

TRAP THE UFOS: MULTIPLY BY 4
YOU NEED: COUNTERS

DIRECTIONS: On your turn, solve a problem. Put a counter on the problem. The last player who puts a counter around a UFO traps it! Put a counter on that UFO. See who can trap the most UFOs.

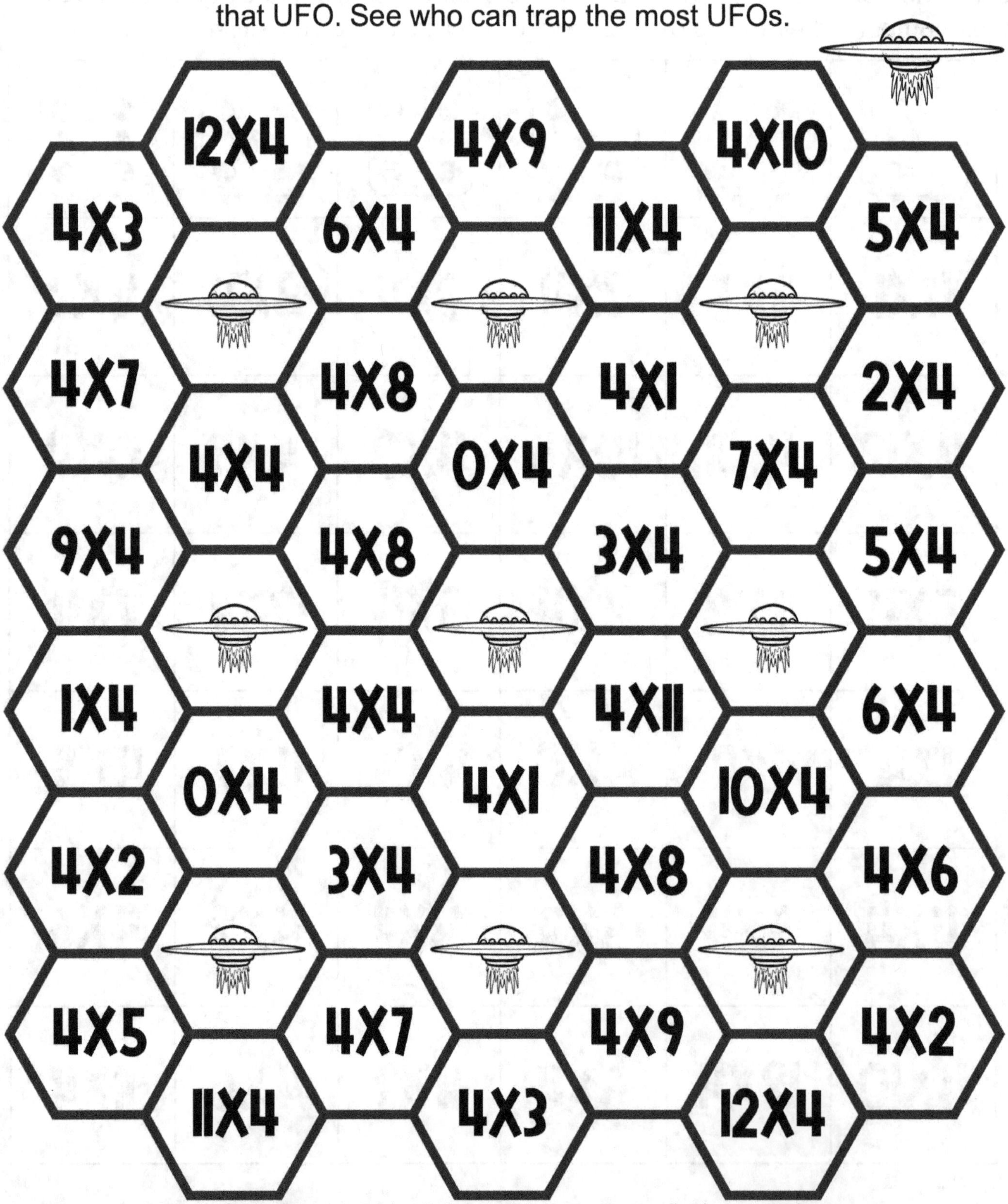

	12X4		4X9		4X10	
4X3		6X4		11X4		5X4
4X7		4X8		4X1		2X4
	4X4		0X4		7X4	
9X4		4X8		3X4		5X4
1X4		4X4		4X11		6X4
	0X4		4X1		10X4	
4X2		3X4		4X8		4X6
4X5		4X7		4X9		4X2
	11X4		4X3		12X4	

© Laura Putman, Bright Minds Engaged, 2024-present, All rights reserved.

TIC-TAC-TOE: MULTIPLY BY 4

DIRECTIONS: Play a game of tic-tac-toe! Before you mark a space as yours, you must solve the problem in that space.

0X4	12X4	4X1
4X7	4X9	11X4
4X10	4X8	4X5

6X4	7X4	2X4
0X4	8X4	4X3
9X4	1X4	4X4

4X9	5X4	4X10
4X1	4X8	4X2
3X4	4X7	4X12

4X4	4X3	12X4
5X4	4X12	4X11
7X4	1X4	8X4

4X7	0X4	4X5
4X6	10X4	4X4
4X2	4X11	3X4

4X6	4X3	4X4
11X4	10X4	4X0
2X4	6X4	9X4

SPIN A PROBLEM: MULTIPLY BY 4
YOU NEED: PAPERCLIP 📎 PENCIL ✏️ CRAYONS 🖍️

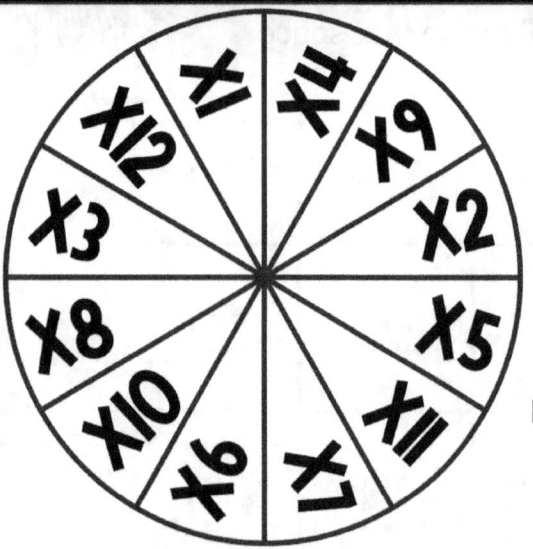

DIRECTIONS: Use a paperclip and pencil to make a spinner. On your turn, spin the paperclip. Multiply that number by 4 and color the product in the table. Each player uses a different color. If the product is not open, your turn is over. See who can solve the most problems!

4	28	48	16	8	32
16	36	24	20	40	12
44	8	16	32	4	16
24	48	20	40	36	44
32	12	36	8	28	24

PIG IN A PEN: MULTIPLY BY 4
YOU NEED: PAPERCLIP 📎 PENCIL ✏️

DIRECTIONS: On their turn, each player spins a number. Multiply the number by 4 and say the answer. If the answer is correct, draw a line to connect 2 dots. When a player completes a box, they write their initial in the box. At the end of the game, boxes are worth 1 point, and boxes with a pig in them are worth 5 points!

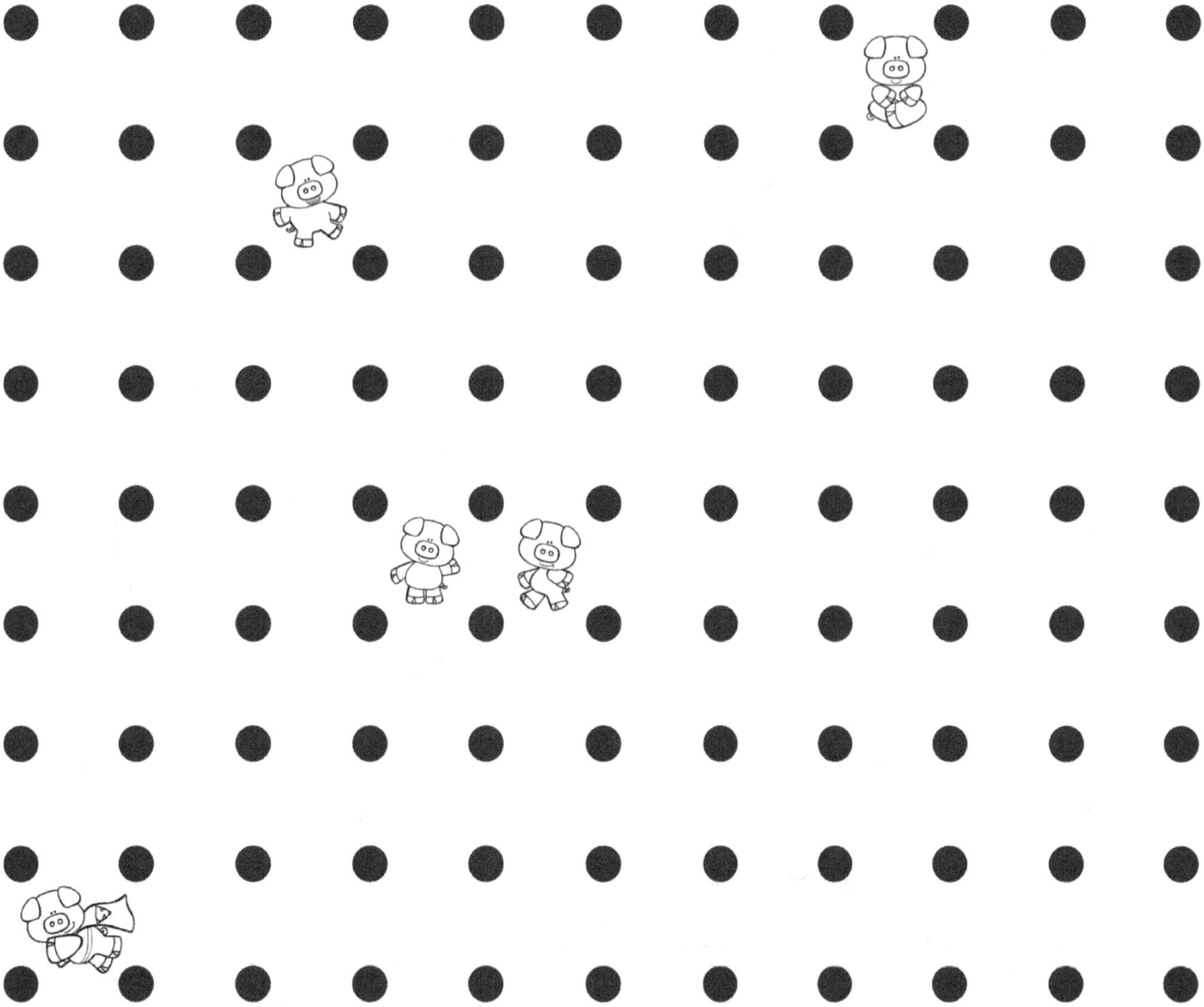

ARRAYS? HOORAY! MULTIPLY BY 4
YOU NEED: CRAYONS 2 DICE

DIRECTIONS: On their turn, each player rolls both dice. Add the dice together and multiply by 4. Draw an array for the problem. Each player uses a different color to make their arrays. Write the multiplication problem inside the array. If your array won't fit, your turn is over. When no more arrays can be made, the game is over. Whoever makes the most arrays, wins!

MULTIPLES OF 4 HUNT
YOU NEED: CRAYONS

Help the kids make it to the ice cream. Count by 4's to follow a path. If you get to 48, start at 4 again. Color or dab the spaces until you get to the bottom.

START →

					4	15	12	81	21	38	56	75
					8	10	20	40	24	18	75	42
					22	12	23	25	32	5	80	79
					16	65	56	40	35	36	52	9
24	15	80	66	27	20	44	10	32	71	40	78
64	14	74	16	28	80	24	28	54	49	44	19
47	37	24	96	32	24	30	55	12	60	4	48
20	48	28	36	2	5	20	16	65	8	7	38
66	23	25	4	40	10	19	72				
41	21	30	18	56	44	48	6				
28	64	81	35	45	40	4	75				
78	35	42	40	71	78	55	8	END →			

FOUR PROBLEMS IN A ROW: MULTIPLY BY 4

YOU NEED: CRAYONS OR COUNTERS

DIRECTIONS: On your turn, solve a problem and color it or cover it. Each player uses a different color. The first player to get 4 in a row wins!

2 X 4	0 X 4	9 X 4	6 X 4	4 X 1	4 X 7
5 X 4	10 X 4	12 X 4	4 X 4	4 X 8	4 X 2
7 X 4	4 X 8	4 X 0	3 X 4	4 X 4	1 X 4
6 X 4	4 X 11	4 X 7	6 X 4	1 X 4	11 X 4
4 X 2	5 X 4	9 X 4	4 X 8	10 X 4	4 X 3
4 X 12	11 X 4	6 X 4	9 X 4	4 X 5	4 X 4

MULTIPLICATION MYSTERY X4'S

YOU NEED: 1 DIE COUNTERS

DIRECTIONS: On your turn, roll a die. Move that number of spaces and solve the problem on the space. If your answer is incorrect, go back to where you started. The first player to the end, wins!

START

4X3		4X0	8X4	4X7		4X9	4X6
7X4		9X4		10X4		3X4	**END**
4X11		4X5		4X11		4X10	
4X4		2X4		2X4		4X4	
4X6		4X8		4X0		4X1	
10X4		12X4		1X4		7X4	
4X12		4X9		4X6		4X11	
1X4	5X4	4X4		8X4	3X4	4X2	

© Laura Putman, Bright Minds Engaged, 2024-present, All rights reserved.

ROLL & SOLVE: MULTIPLY BY 5

YOU NEED: 1 DIE CRAYONS

DIRECTIONS: Assign one player even numbers on the die and the other player odd numbers. Take turns rolling. If a player rolls one of their numbers on the die, they solve the next problem under that die and color the space. If they do not roll one of their numbers, their turn is over. See who can fill their columns first!

⚀	⚁	⚂	⚃	⚄	⚅
5X1	5X5	7X5	5X10	8X5	2X5
5X12	5X11	10X5	5X8	3X5	0X5
5X5	5X8	5X4	5X3	12X5	7X5
5X2	1X5	9X5	4X5	5X7	5X9
4X5	5X2	5X5	11X5	2X5	5X6
5X0	12X5	6X5	5X3	5X6	11X5

TRAP THE SHARKS: MULTIPLY BY 5

YOU NEED: COUNTERS

DIRECTIONS: On your turn, solve a problem. Put a counter on the problem. The last player who puts a counter around a shark traps it! Put a counter on that shark. See who can trap the most sharks.

12X5 5X9 5X10

5X3 6X5 11X5 5X5

5X7 5X8 5X1 2X5

5X4 0X5 7X5

9X5 5X8 3X5 5X5

1X5 4X5 5X11 6X5

0X5 5X1 10X5

5X2 3X5 5X8 5X6

5X5 5X7 5X9 5X2

11X5 5X3 12X5

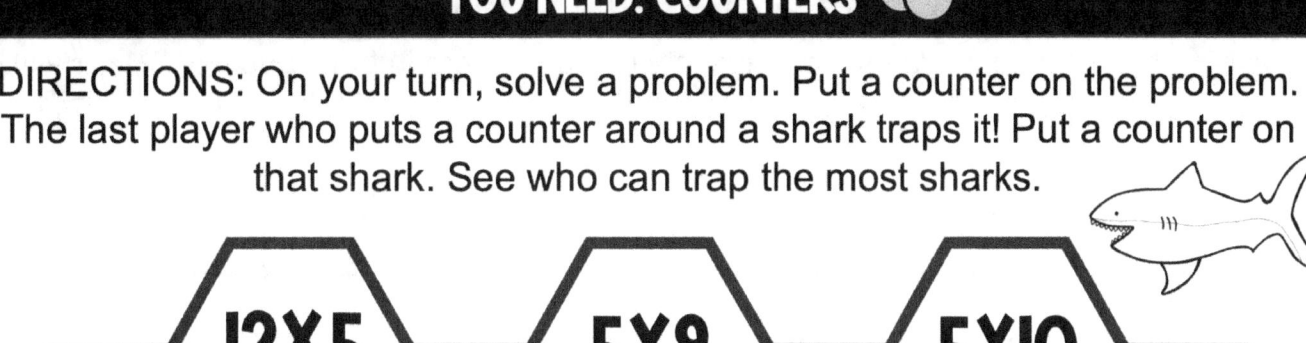

TIC-TAC-TOE: MULTIPLY BY 5

DIRECTIONS: Play a game of tic-tac-toe! Before you mark a space as yours, you must solve the problem in that space.

0X5	12X5	5X1
5X7	5X9	11X5
5X10	5X8	5X5

6X5	7X5	2X5
0X5	8X5	5X3
9X5	1X5	5X4

5X9	5X5	5X10
5X1	5X8	5X2
3X5	5X7	5X12

4X5	5X3	12X5
5X5	5X12	5X11
7X5	1X5	8X5

5X7	0X5	5X5
5X6	10X5	4X5
5X2	5X11	3X5

5X6	5X3	5X4
11X5	10X5	5X0
2X5	6X5	9X5

SPIN A PROBLEM: MULTIPLY BY 5
YOU NEED: PAPERCLIP PENCIL CRAYONS

DIRECTIONS: Use a paperclip and pencil to make a spinner. On your turn, spin the paperclip. Multiply that number by 5 and color the product in the table. Each player uses a different color. If the product is not open, your turn is over. See who can solve the most problems!

5	35	60	20	10	40
20	45	30	25	50	15
55	10	15	40	5	20
30	60	25	50	45	55
25	15	45	10	35	30

PIG IN A PEN: MULTIPLY BY 5
YOU NEED: PAPERCLIP 📎 PENCIL ✏️

DIRECTIONS: On their turn, each player spins a number. Multiply the number by 5 and say the answer. If the answer is correct, draw a line to connect 2 dots. When a player completes a box, they write their initial in the box. At the end of the game, boxes are worth 1 point, and boxes with a pig in them are worth 5 points!

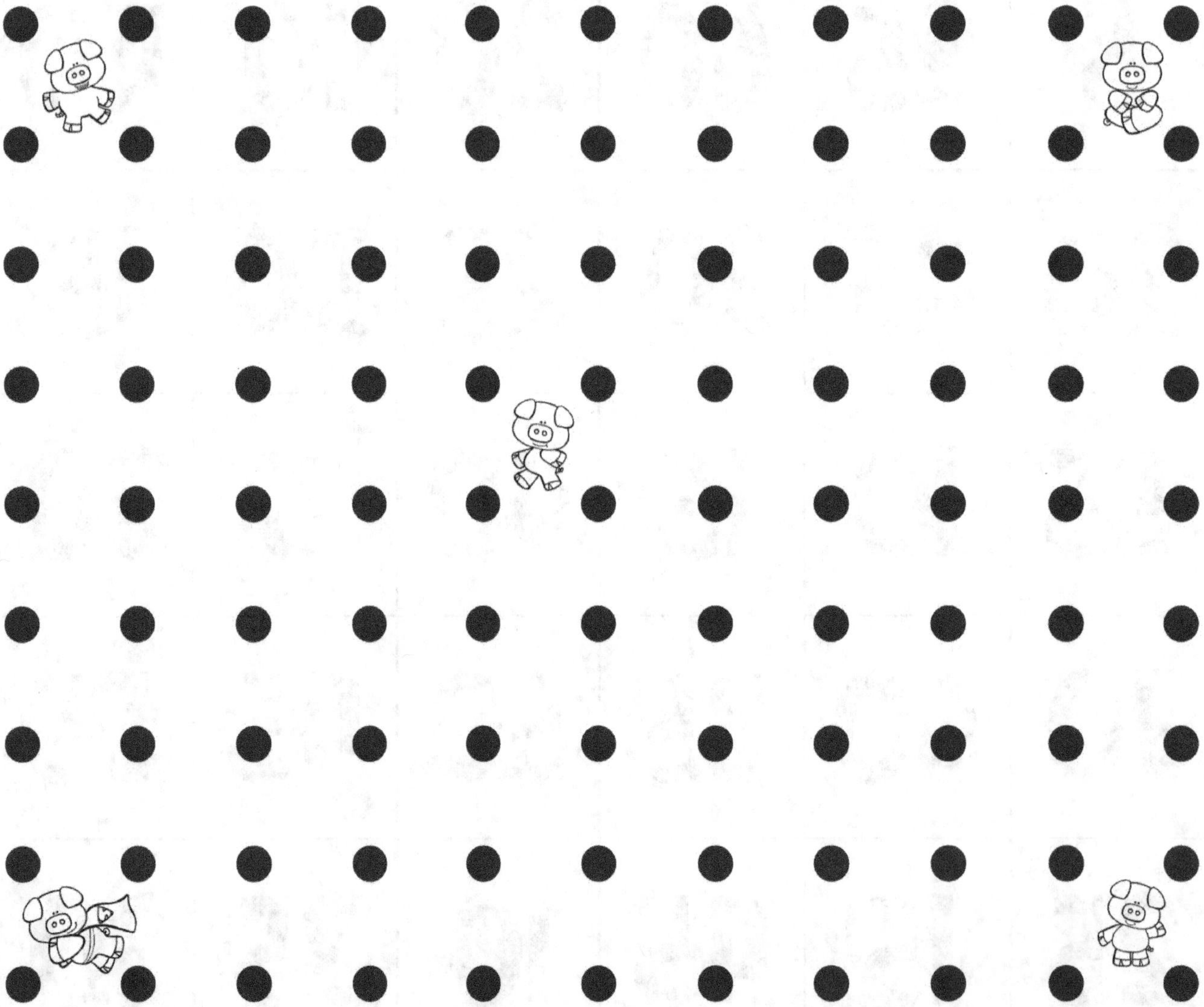

ARRAYS? HOORAY! MULTIPLY BY 5

YOU NEED: CRAYONS · 2 DICE

DIRECTIONS: On their turn, each player rolls both dice. Add the dice together and multiply by 5. Draw an array for the problem. Each player uses a different color to make their arrays. Write the multiplication problem inside the array. If your array won't fit, your turn is over. When no more arrays can be made, the game is over. Whoever makes the most arrays, wins!

MULTIPLES OF 5 HUNT
YOU NEED: CRAYONS

Help the panda find its friends. Count by 5's to follow a path. If you get to 60, start at 5 again. Color or dab the spaces until you get to the bottom.

			START →	5	15	12	81	21	38	56	25
				12	10	20	40	24	63	75	42
				22	16	23	25	30	5	80	79
				6	65	56	40	35	36	52	9
24	15	80	66	27	44	45	10	42	71	8	78
64	10	74	16	88	80	50	35	54	49	62	13
47	37	24	96	84	41	16	55	56	60	18	40
20	48	28	20	83	5	60	63	65	27	7	38
66	23	25	40	15	10	19	72				
41	21	30	18	56	63	74	18				
28	52	81	35	45	50	72	75				
78	35	42	40	71	78	55	60	→ END			

FOUR PROBLEMS IN A ROW: MULTIPLY BY 5
YOU NEED: CRAYONS

DIRECTIONS: On your turn, solve a problem and color it or cover it. Each player uses a different color. The first player to get 4 in a row wins!

2 X 5	0 X 5	9 X 5	6 X 5	5 X 1	5 X 7
5 X 5	10 X 5	12 X 5	5 X 4	5 X 8	5 X 2
7 X 5	5 X 8	5 X 0	3 X 5	4 X 5	1 X 5
6 X 5	5 X 11	5 X 7	6 X 5	1 X 5	11 X 5
5 X 2	5 X 5	9 X 5	5 X 8	10 X 5	5 X 3
5 X 12	11 X 5	6 X 5	9 X 5	5 X 5	4 X 5

POPPIN' PRODUCTS X5'S

YOU NEED: 1 DIE COUNTERS

DIRECTIONS: On your turn, roll a die. Move that number of spaces and solve the problem on the space. If your answer is incorrect, go back to where you started. The first player to the end, wins!

START

| 5X5 | 6X5 | 5X4 | 7X5 | 12X5 | 5X0 | 2X5 | 5X9 |

| | | | | | | | 8X5 |

| 2X5 | 5X4 | 6X5 | 5X8 | 3X5 | 5X1 | | 5X3 |

| 5X10 | | | **END** | | 5X10 | | 10X5 |

| 1X5 | | 5X7 | | 5X7 | | | 1X5 |

| 5X2 | | 5X9 | 5X0 | 11X5 | 5X5 | | 11X5 |

| 5X11 | | | | | | | 5X5 |

| 8X5 | 5X7 | 5X12 | 0X5 | 5X6 | 5X9 | 5X3 | 4X5 |

© Laura Putman, Bright Minds Engaged, 2024-present, All rights reserved.

ROLL & SOLVE: MULTIPLY BY 6
YOU NEED: 1 DIE  CRAYONS

DIRECTIONS: Assign one player even numbers on the die and the other player odd numbers. Take turns rolling. If a player rolls one of their numbers on the die, they solve the next problem under that die and color the space. If they do not roll one of their numbers, their turn is over. See who can fill their columns first!

⚀	⚁	⚂	⚃	⚄	⚅
6X1	6X5	7X6	2X6	8X6	6X3
6X12	6X11	10X6	6X8	6X9	0X6
5X6	6X8	4X6	3X6	12X6	7X6
1X6	1X6	6X0	6X4	6X7	6X9
6X4	9X6	5X6	6X11	6X2	6X6
6X10	12X6	6X6	6X3	6X6	11X6

TRAP THE CREATURES: MULTIPLY BY 6
YOU NEED: COUNTERS

DIRECTIONS: On your turn, solve a problem. Put a counter on the problem. The last player who puts a counter around a creature traps it! Put a counter on that creature. See who can trap the most creatures.

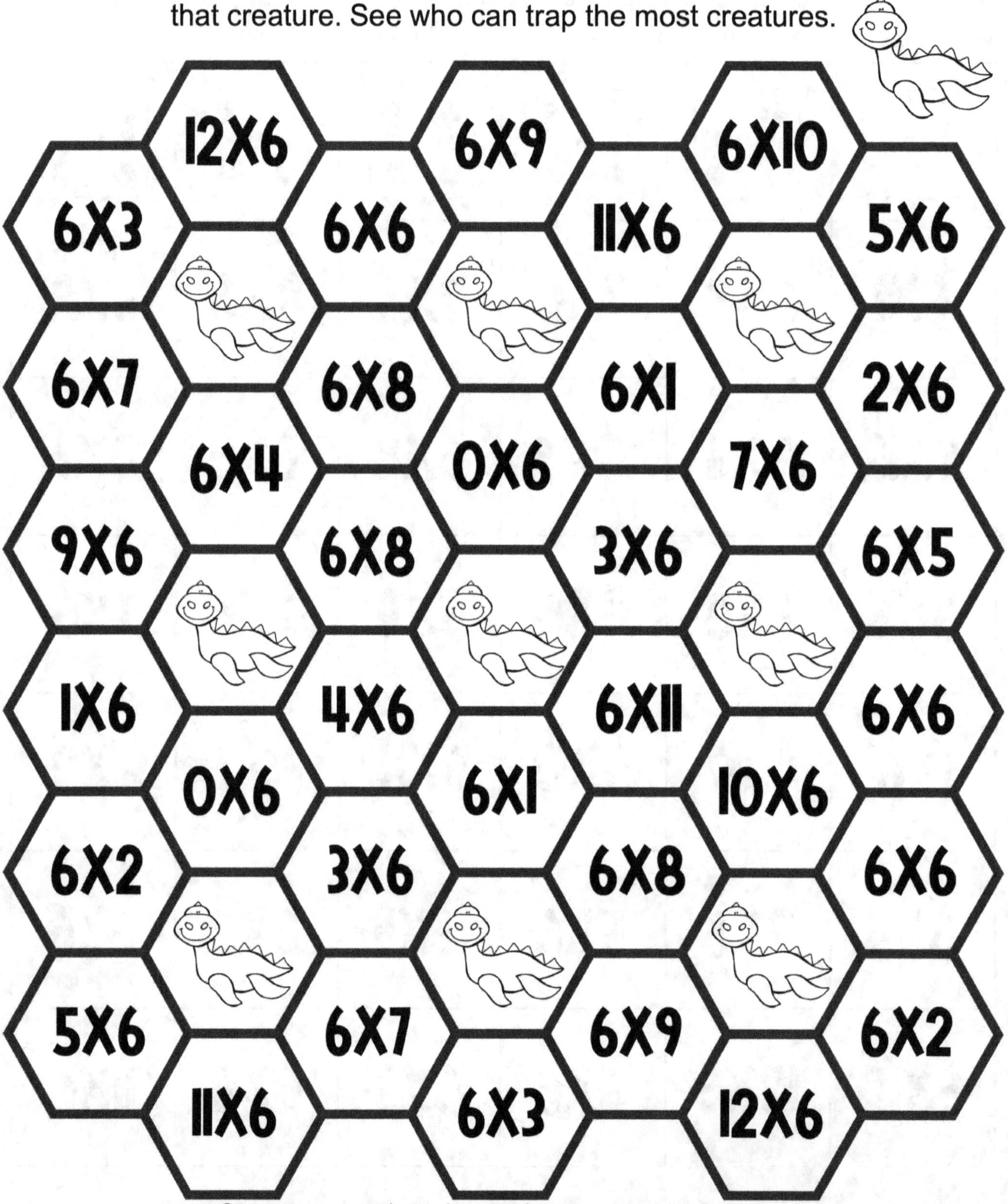

12X6	6X9	6X10
6X3	6X6	11X6	5X6
6X7	6X8	6X11	2X6
6X4	0X6	7X6
9X6	6X8	3X6	6X5
1X6	4X6	6X11	6X6
0X6	6X11	10X6
6X2	3X6	6X8	6X6
5X6	6X7	6X9	6X2
11X6	6X3	12X6

TIC-TAC-TOE: MULTIPLY BY 6

DIRECTIONS: Play a game of tic-tac-toe! Before you mark a space as yours, you must solve the problem in that space.

0X6	12X6	6X1
6X7	6X9	11X6
6X10	6X8	6X5

6X6	7X6	2X6
0X6	8X6	6X3
9X6	1X6	6X4

6X9	5X6	6X10
6X1	6X8	6X2
3X6	6X7	6X12

4X6	6X3	12X6
5X6	6X12	6X11
7X6	1X6	8X6

6X7	0X6	6X5
6X6	10X6	4X6
6X2	6X11	3X6

6X6	6X3	6X4
11X6	10X6	6X0
2X6	6X6	9X6

SPIN A PROBLEM: MULTIPLY BY 6
YOU NEED: PAPERCLIP 📎 PENCIL ✏️ CRAYONS 🖍️

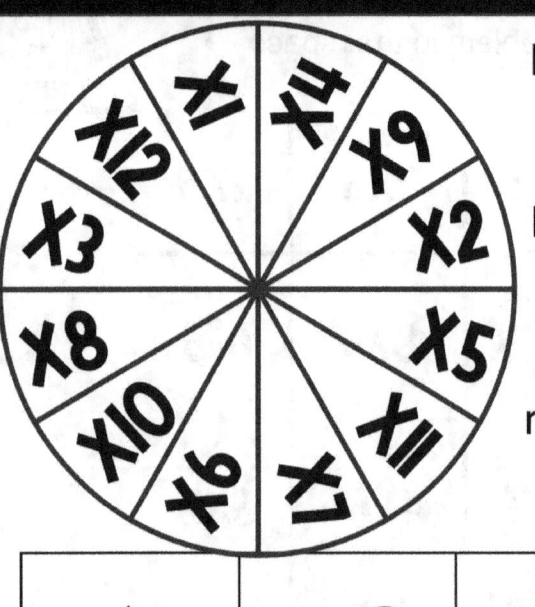

DIRECTIONS: Use a paperclip and pencil to make a spinner. On your turn, spin the paperclip. Multiply that number by 6 and color the product in the table. Each player uses a different color. If the product is not open, your turn is over. See who can solve the most problems!

6	42	72	24	12	48
24	54	36	30	60	18
66	12	18	48	6	24
36	72	30	60	54	66
30	18	54	12	42	36

© Laura Putman, Bright Minds Engaged, 2024-present, All rights reserved.

PIG IN A PEN: MULTIPLY BY 6
YOU NEED: PAPERCLIP 📎 PENCIL ✏️

DIRECTIONS: On their turn, each player spins a number. Multiply the number by 6 and say the answer. If the answer is correct, draw a line to connect 2 dots. When a player completes a box, they write their initial in the box. At the end of the game, boxes are worth 1 point, and boxes with a pig in them are worth 5 points!

ARRAYS? HOORAY! MULTIPLY BY 6
YOU NEED: CRAYONS   2 DICE

DIRECTIONS: On their turn, each player rolls both dice. Add the dice together and multiply by 6. Draw an array for the problem. Each player uses a different color to make their arrays. Write the multiplication problem inside the array. If your array won't fit, your turn is over. When no more arrays can be made, the game is over. Whoever makes the most arrays, wins!

MULTIPLES OF 6 HUNT
YOU NEED: CRAYONS

Help the raccoon win the snowball fight! Count by 6's to follow a path. If you get to 72, start at 6 again. Color or dab the spaces until you get to the bottom.

				6	15	12	18	21	38	56	33
				12	14	20	40	24	63	12	42
				36	18	24	28	32	9	80	79
				6	65	22	30	35	36	52	9
24	15	80	6	27	44	33	36	42	71	8	78
64	10	74	16	8	18	72	35	54	48	62	13
47	37	24	60	40	41	5	60	54	6	88	40
21	48	28	7	32	72	66	63	11	27	42	38
18	23	18	40	6	32	19	72				
41	21	79	18	56	12	18	24				
28	52	81	49	15	70	30	75				
78	35	42	35	71	78	77	36				

FOUR PROBLEMS IN A ROW: MULTIPLY BY 6
YOU NEED: CRAYONS OR COUNTERS

DIRECTIONS: On your turn, solve a problem and color it or cover it. Each player uses a different color. The first player to get 4 in a row wins!

2 X 6	0 X 6	9 X 6	6 X 6	6 X 1	6 X 7
6 X 5	10 X 6	12 X 6	6 X 4	6 X 8	6 X 2
7 X 6	6 X 8	6 X 0	3 X 6	4 X 6	1 X 6
6 X 6	6 X 11	6 X 7	6 X 6	1 X 6	11 X 6
6 X 2	5 X 6	9 X 6	6 X 8	10 X 6	6 X 3
6 X 12	11 X 6	6 X 6	9 X 6	6 X 5	4 X 6

© Laura Putman, Bright Minds Engaged, 2024-present, All rights reserved.

MATH BATTLE: X6'S
YOU NEED: 1 DIE 🎲 COUNTERS ⬤

DIRECTIONS: On your turn, roll a die. Move that number of spaces and solve the problem on the space. If your answer is incorrect, go back to where you started. The first player to the end, wins!

START

6X3		6X0	8X6	6X7		6X9	6X6
7X6		9X6		10X6		3X6	**END**
6X12		6X5		6X12		6X10	
4X6		6X2		2X6		4X6	
6X6		6X8		6X0		6X1	
10X6		12X6		1X6		7X6	
6X12		6X9		6X6		6X11	
1X6	5X6	4X6		8X6	3X6	6X2	

© Laura Putman, Bright Minds Engaged, 2024-present, All rights reserved.

ROLL & SOLVE: MULTIPLY BY 7

YOU NEED: 1 DIE • CRAYONS

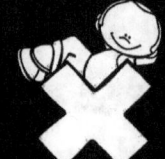

DIRECTIONS: Assign one player even numbers on the die and the other player odd numbers. Take turns rolling. If a player rolls one of their numbers on the die, they solve the next problem under that die and color the space. If they do not roll one of their numbers, their turn is over. See who can fill their columns first!

⚀	⚁	⚂	⚃	⚄	⚅
7X1	5X7	7X7	7X10	8X7	2X7
7X12	7X11	10X7	7X8	3X7	0X7
5X7	7X8	7X4	7X3	12X7	7X7
7X2	1X7	9X7	4X7	7X7	7X9
4X7	7X2	7X5	11X7	2X7	7X6
7X0	12X7	6X7	7X3	7X6	11X7

TRAP THE TREASURE: MULTIPLY BY 7
YOU NEED: COUNTERS

DIRECTIONS: On your turn, solve a problem. Put a counter on the problem. The last player who puts a counter around a chest traps it! Put a counter on that chest. See who can trap the most chests.

Board problems (by row):
- 12X7, 7X9, 7X10
- 7X3, 6X7, 11X7, 5X7
- 7X7, 7X8, 7X1, 2X7
- 7X4, 0X7, 7X7
- 9X7, 7X8, 3X7, 7X5
- 1X7, 4X7, 7X11, 6X7
- 0X7, 7X1, 10X7
- 7X2, 3X7, 7X8, 7X6
- 5X7, 7X7, 7X9, 7X2
- 11X7, 7X3, 12X7

© Laura Putman, Bright Minds Engaged, 2024-present, All rights reserved.

55

TIC-TAC-TOE: MULTIPLY BY 7

DIRECTIONS: Play a game of tic-tac-toe! Before you mark a space as yours, you must solve the problem in that space.

0X7	12X7	7X1		7X6	7X7	2X7
7X7	7X9	11X7		0X7	8X7	7X3
7X10	7X8	7X5		9X7	1X7	7X4

7X9	5X7	7X10		4X7	7X3	12X7
7X1	7X8	7X2		5X7	7X12	7X11
3X7	7X7	7X12		7X7	1X7	8X7

7X7	0X7	7X5		6X7	7X3	7X4
6X7	10X7	4X7		11X7	10X7	7X0
7X2	7X11	3X7		2X7	7X6	9X7

SPIN A PROBLEM: MULTIPLY BY 7
YOU NEED: PAPERCLIP 📎 PENCIL ✏️ CRAYONS

DIRECTIONS: Use a paperclip and pencil to make a spinner. On your turn, spin the paperclip. Multiply that number by 7 and color the product in the table. Each player uses a different color. If the product is not open, your turn is over. See who can solve the most problems!

7	49	84	28	14	56
28	63	42	35	70	21
77	14	21	56	7	28
42	84	35	70	63	77
35	21	63	14	49	42

© Laura Putman, Bright Minds Engaged, 2024-present. All rights reserved.

PIG IN A PEN: MULTIPLY BY 7
YOU NEED: PAPERCLIP ✏ PENCIL ✏

DIRECTIONS: On their turn, each player spins a number. Multiply the number by 7 and say the answer. If the answer is correct, draw a line to connect 2 dots. When a player completes a box, they write their initial in the box. At the end of the game, boxes are worth 1 point, and boxes with a pig in them are worth 5 points!

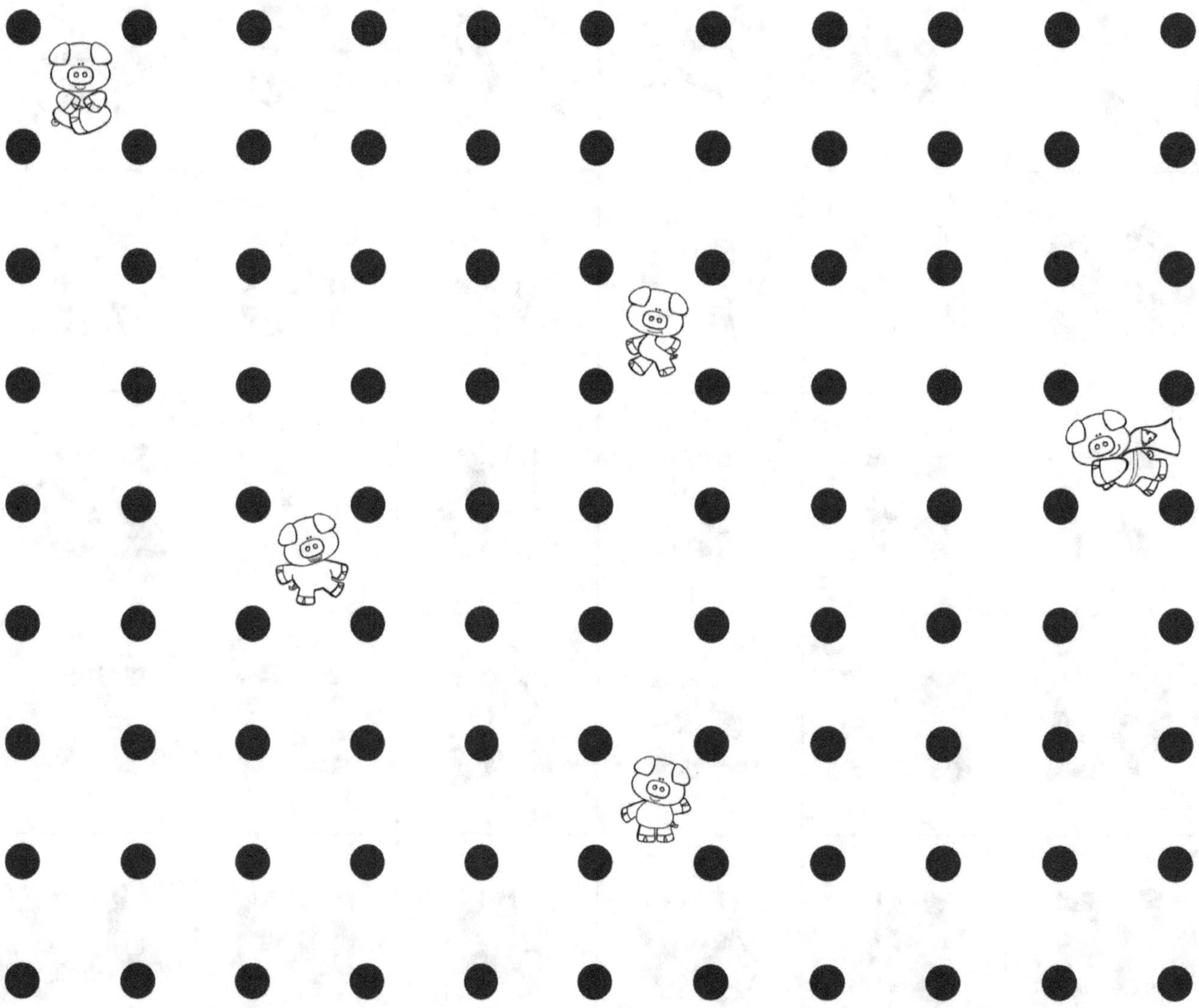

ARRAYS? HOORAY! MULTIPLY BY 7
YOU NEED: CRAYONS 2 DICE

DIRECTIONS: On their turn, each player rolls both dice. Add the dice together and multiply by 7. Draw an array for the problem. Each player uses a different color to make their arrays. Write the multiplication problem inside the array. If your array won't fit, your turn is over. When no more arrays can be made, the game is over. Whoever makes the most arrays, wins!

MULTIPLES OF 7 HUNT
YOU NEED: CRAYONS

Help the kids win the water fight! Count by 7's to follow a path. If you get to 84, start at 7 again. Color or dab the spaces until you get to the bottom.

			7	15	12	81	21	38	56	48
			16	14	21	40	24	63	75	42
			36	24	23	28	32	9	80	79
			6	65	56	40	35	36	52	9
24	15	80	66	27	44	64	13	42	71	8	78
64	10	74	16	88	80	72	35	54	49	62	13
47	37	24	96	84	41	16	52	56	66	88	40
21	48	28	7	83	77	70	63	11	27	78	38
66	23	14	40	48	32	19	72				
41	21	79	18	56	63	74	18				
28	52	81	49	15	70	72	75				
78	35	42	35	71	78	77	84	END			

FOUR PROBLEMS IN A ROW: MULTIPLY BY 7
YOU NEED: CRAYONS OR COUNTERS

DIRECTIONS: On your turn, solve a problem and color it or cover it. Each player uses a different color. The first player to get 4 in a row wins!

2 X 7	0 X 7	9 X 7	6 X 7	7 X 1	7 X 7
7 X 5	10 X 7	12 X 7	7 X 4	7 X 8	7 X 2
7 X 7	7 X 8	7 X 0	3 X 7	4 X 7	1 X 7
6 X 7	7 X 11	7 X 7	7 X 6	1 X 7	11 X 7
7 X 2	5 X 7	9 X 7	7 X 8	10 X 7	7 X 3
7 X 12	11 X 7	7 X 6	9 X 7	7 X 5	4 X 7

© Laura Putman, Bright Minds Engaged, 2024-present, All rights reserved.

FOOTBALL FACTS X7'S

YOU NEED: 1 DIE COUNTERS

DIRECTIONS: On your turn, roll a die. Move that number of spaces and solve the problem on the space. If your answer is incorrect, go back to where you started. The first player to the end, wins!

START

| 5X7 | 6X7 | 7X4 | 7X7 | 12X7 | 7X0 | 2X7 | 7X9 |

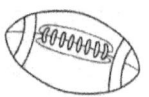    8X7

| 2X7 | 7X4 | 6X7 | 7X8 | 3X7 | 7X1 | | 7X3 |

| 7X10 | | | | | 7X10 | | 10X7 |
| 1X7 | 7X7 END | | | | 4X7 | | 1X7 |

| 7X2 | | 7X9 | 7X0 | 11X7 | 5X7 | | 11X7 |

| 7X11 | | | | | | | 5X7 |
| 8X7 | 7X7 | 7X12 | 0X7 | 7X6 | 7X9 | 7X3 | 4X7 |

© Laura Putman, Bright Minds Engaged, 2024-present, All rights reserved.

ROLL & SOLVE: MULTIPLY BY 8
YOU NEED: 1 DIE CRAYONS

DIRECTIONS: Assign one player even numbers on the die and the other player odd numbers. Take turns rolling. If a player rolls one of their numbers on the die, they solve the next problem under that die and color the space. If they do not roll one of their numbers, their turn is over. See who can fill their columns first!

⚀	⚁	⚂	⚃	⚄	⚅
8X1	8X5	7X8	2X8	8X8	8X3
8X12	8X11	10X8	8X8	8X9	0X8
5X8	8X8	4X8	3X8	12X8	7X8
1X8	1X8	8X0	8X4	8X7	8X9
8X4	9X8	5X8	8X11	8X2	8X6
8X10	12X8	8X6	8X3	6X8	11X8

TRAP THE MONKEYS: MULTIPLY BY 8
YOU NEED: COUNTERS

DIRECTIONS: On your turn, solve a problem. Put a counter on the problem. The last player who puts a counter around a monkey traps it! Put a counter on that monkey. See who can trap the most monkeys.

- 12X8
- 8X9
- 8X10
- 8X3
- 6X8
- 11X8
- 5X8
- 7X8
- 8X5
- 8X1
- 2X8
- 8X4
- 0X8
- 8X7
- 9X8
- 8X8
- 3X8
- 8X5
- 1X8
- 4X8
- 8X11
- 6X8
- 0X8
- 8X1
- 10X8
- 8X2
- 3X8
- 8X8
- 8X6
- 5X8
- 8X7
- 8X9
- 8X2
- 11X8
- 8X3
- 12X8

TIC-TAC-TOE: MULTIPLY BY 8

DIRECTIONS: Play a game of tic-tac-toe! Before you mark a space as yours, you must solve the problem in that space.

0X8	12X8	8X1
8X7	8X9	11X8
8X10	8X8	8X5

8X6	7X8	2X8
0X8	8X8	8X3
9X8	1X8	8X4

8X9	5X8	8X10
8X1	8X8	8X2
3X8	8X7	8X12

4X8	8X3	12X8
5X8	8X12	8X11
7X8	1X8	8X8

8X7	0X8	8X5
6X8	10X8	4X8
8X2	8X11	3X8

6X8	8X3	8X4
11X8	10X8	8X0
2X8	8X6	8X7

SPIN A PROBLEM: MULTIPLY BY 8

YOU NEED: PAPERCLIP 📎 PENCIL ✏️ CRAYONS

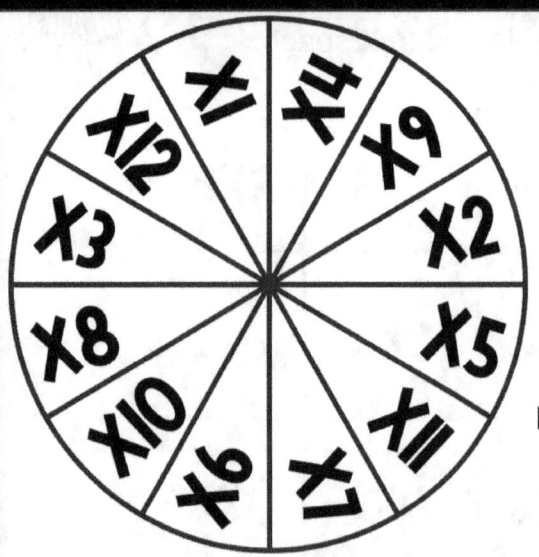

DIRECTIONS: Use a paperclip and pencil to make a spinner. On your turn, spin the paperclip. Multiply that number by 8 and color the product in the table. Each player uses a different color. If the product is not open, your turn is over. See who can solve the most problems!

8	56	96	32	16	64
32	72	48	40	80	24
88	16	24	64	8	64
48	96	40	80	72	88
40	24	72	16	56	48

© Laura Putman, Bright Minds Engaged, 2024-present, All rights reserved.

PIG IN A PEN: MULTIPLY BY 8
YOU NEED: PAPERCLIP 📎 PENCIL ✏️

DIRECTIONS: On their turn, each player spins a number. Multiply the number by 8 and say the answer. If the answer is correct, draw a line to connect 2 dots. When a player completes a box, they write their initial in the box. At the end of the game, boxes are worth 1 point, and boxes with a pig in them are worth 5 points!

© Laura Putman, Bright Minds Engaged, 2024-present, All rights reserved.

ARRAYS? HOORAY! MULTIPLY BY 8

YOU NEED: CRAYONS   2 DICE

DIRECTIONS: On their turn, each player rolls both dice. Add the dice together and multiply by 8. Draw an array for the problem. Each player uses a different color to make their arrays. Write the multiplication problem inside the array. If your array won't fit, your turn is over. When no more arrays can be made, the game is over. Whoever makes the most arrays, wins!

MULTIPLES OF 8 HUNT
YOU NEED: CRAYONS

Help the kid get to the candy! Count by 8's to follow a path. If you get to 96, start at 8 again. Color or dab the spaces until you get to the bottom.

		START →	8	16	12	81	21	38	56	55
			18	14	24	40	24	63	75	42
			36	24	23	32	38	9	80	79
			6	65	56	42	40	36	52	9
24	48	80	66	27	44	64	13	42	48	8	78
63	56	40	44	88	80	72	35	54	49	56	13
64	72	24	32	84	16	19	96	56	66	88	64
21	88	80	7	24	77	8	63	88	80	72	38
96	94	14	40	48	36	19	72				
8	24	79	18	63	77	74	18				
16	52	32	56	64	80	89	94				
78	40	48	35	71	78	88	96	END →			

FOUR PROBLEMS IN A ROW: MULTIPLY BY 8

YOU NEED: CRAYONS OR COUNTERS

DIRECTIONS: On your turn, solve a problem and color it or cover it. Each player uses a different color. The first player to get 4 in a row wins!

2 X 8	0 X 8	9 X 8	6 X 8	8 X 1	8 X 7
8 X 5	10 X 8	12 X 8	8 X 4	8 X 8	8 X 2
7 X 8	8 X 8	8 X 0	3 X 8	4 X 8	1 X 8
6 X 8	8 X 11	8 X 7	8 X 6	1 X 8	11 X 8
8 X 2	5 X 8	9 X 8	8 X 8	10 X 8	8 X 3
8 X 12	11 X 8	8 X 6	9 X 8	8 X 5	4 X 8

MULTIPLICATION MYSTERY X8'S

YOU NEED: 1 DIE COUNTERS

DIRECTIONS: On your turn, roll a die. Move that number of spaces and solve the problem on the space. If your answer is incorrect, go back to where you started. The first player to the end, wins!

START

8X3		8X0	8X8	7X8		8X9	8X6
7X8		9X8		10X8		3X8	**END**
8X11		8X5		8X11		8X10	
4X8		2X8		2X8		4X8	
8X6		8X8		8X0		8X1	
10X8		12X8		1X8		7X8	
8X12		8X9		8X6		8X11	
1X8	5X8	8X4		8X8	3X8	8X2	

© Laura Putman, Bright Minds Engaged, 2024-present, All rights reserved.

ROLL & SOLVE: MULTIPLY BY 9

YOU NEED: 1 DIE 🎲 CRAYONS

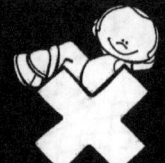

DIRECTIONS: Assign one player even numbers on the die and the other player odd numbers. Take turns rolling. If a player rolls one of their numbers on the die, they solve the next problem under that die and color the space. If they do not roll one of their numbers, their turn is over. See who can fill their columns first!

⚀	⚁	⚂	⚃	⚄	⚅
9X1	5X9	9X7	9X10	8X9	2X9
9X12	9X11	10X9	9X8	3X9	0X9
5X9	9X8	9X4	9X3	12X9	9X7
9X2	1X9	9X9	4X9	7X9	9X9
4X9	9X2	9X5	11X9	2X9	9X6
9X0	12X9	6X9	9X3	9X6	11X9

© Laura Putman, Bright Minds Engaged, 2024-present, All rights reserved.

TRAP THE SUBMARINES: MULTIPLY BY 9
YOU NEED: COUNTERS

DIRECTIONS: On your turn, solve a problem. Put a counter on the problem. The last player who puts a counter around a sub traps it! Put a counter on that sub. See who can trap the most subs.

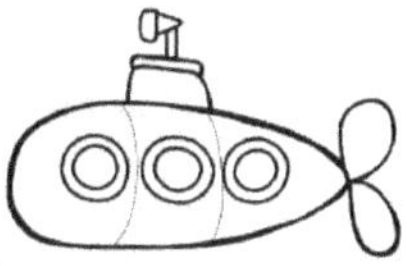

- 12X9
- 9X9
- 9X10
- 9X3
- 6X9
- 11X9
- 5X9
- 7X9
- 9X5
- 9X1
- 2X9
- 9X4
- 0X9
- 9X7
- 9X9
- 8X9
- 3X9
- 9X5
- 1X9
- 4X9
- 9X11
- 6X9
- 0X9
- 9X1
- 10X9
- 9X2
- 3X9
- 9X8
- 9X6
- 5X9
- 9X7
- 9X9
- 9X2
- 11X9
- 9X3
- 12X9

TIC-TAC-TOE: MULTIPLY BY 9

DIRECTIONS: Play a game of tic-tac-toe! Before you mark a space as yours, you must solve the problem in that space.

0X9	12X9	9X1
9X7	9X9	11X9
9X10	8X9	9X5

9X6	7X9	2X9
0X9	9X8	9X3
9X9	1X9	9X4

9X9	5X9	9X10
9X1	9X8	9X2
3X9	9X7	9X12

4X9	9X3	12X9
5X9	9X12	9X11
7X9	1X9	8X9

9X7	0X9	9X5
6X9	10X9	4X9
9X2	9X11	3X9

6X9	9X3	9X4
11X9	10X9	9X0
2X9	9X6	9X7

SPIN A PROBLEM: MULTIPLY BY 9
YOU NEED: PAPERCLIP 🖇 PENCIL ✏ CRAYONS

DIRECTIONS: Use a paperclip and pencil to make a spinner. On your turn, spin the paperclip. Multiply that number by 9 and color the product in the table. Each player uses a different color. If the product is not open, your turn is over. See who can solve the most problems!

9	63	108	36	18	72
36	81	54	45	90	27
99	18	27	72	9	36
54	108	45	90	81	99
45	27	81	18	63	54

PIG IN A PEN: MULTIPLY BY 9
YOU NEED: PAPERCLIP ✏ PENCIL ✏

DIRECTIONS: On their turn, each player spins a number. Multiply the number by 9 and say the answer. If the answer is correct, draw a line to connect 2 dots. When a player completes a box, they write their initial in the box. At the end of the game, boxes are worth 1 point, and boxes with a pig in them are worth 5 points!

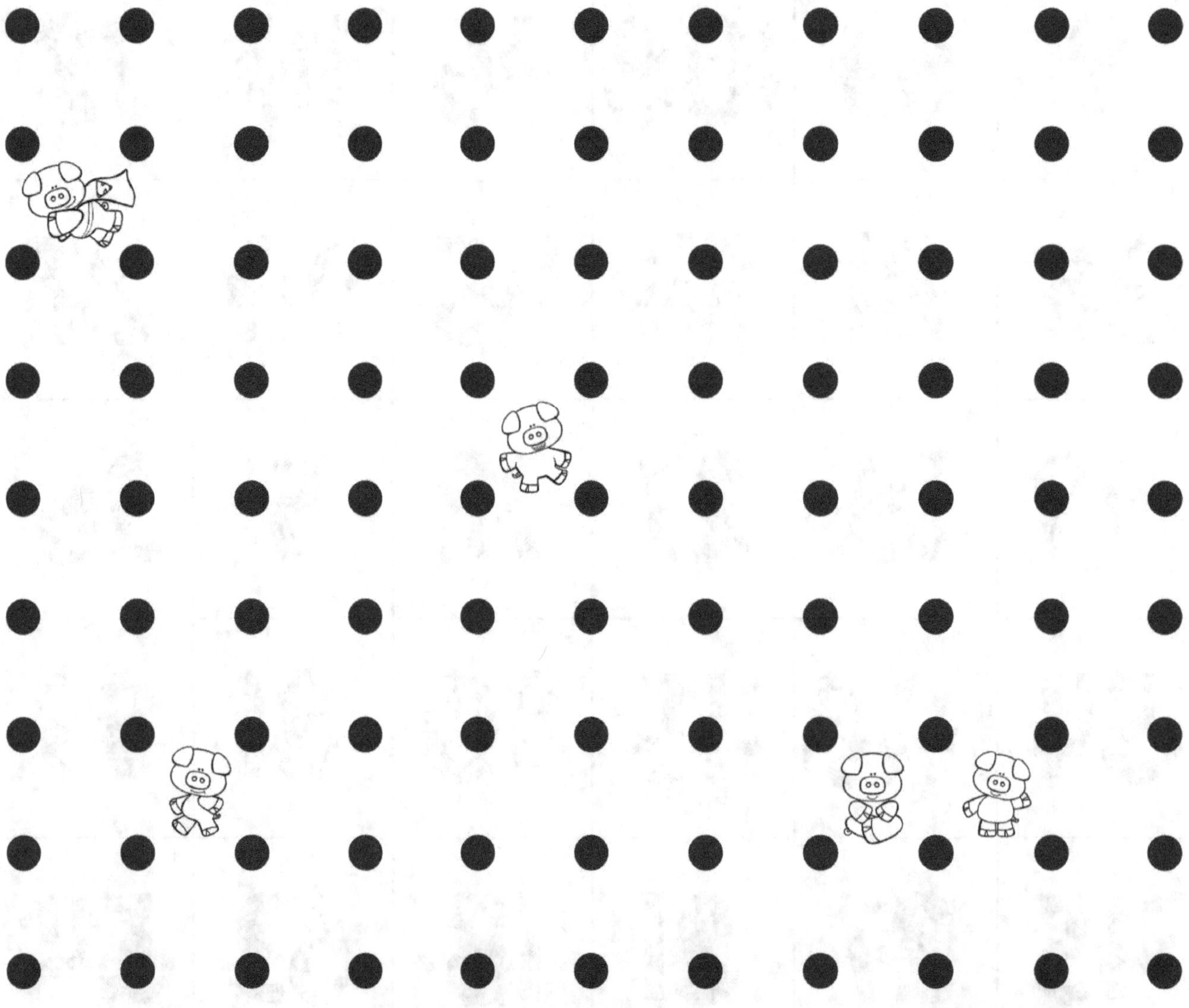

ARRAYS? HOORAY! MULTIPLY BY 9
YOU NEED: CRAYONS 2 DICE

DIRECTIONS: On their turn, each player rolls both dice. Add the dice together and multiply by 9. Draw an array for the problem. Each player uses a different color to make their arrays. Write the multiplication problem inside the array. If your array won't fit, your turn is over. When no more arrays can be made, the game is over. Whoever makes the most arrays, wins!

MULTIPLES OF 9 HUNT
YOU NEED: CRAYONS

Help make a touchdown! Count by 9's to follow a path. If you get to 108, start at 9 again. Color or dab the spaces until you get to the bottom.

			START →	9	16	12	81	21	38	56	19
				18	14	24	40	84	81	92	99
				36	27	23	32	72	9	90	108
				6	45	56	63	74	36	52	9
24	48	80	66	27	44	54	64	42	48	18	19
63	56	40	44	109	80	72	35	54	49	29	27
64	72	24	9	108	99	19	96	56	45	36	29
21	23	18	7	24	77	90	63	54	80	72	38
96	27	34	40	48	36	81	72				
8	36	45	76	72	81	74	18				
16	52	54	63	64	90	99	94				
78	40	48	35	71	78	110	108	→ END			

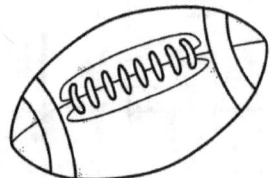 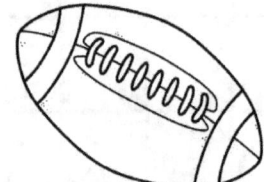 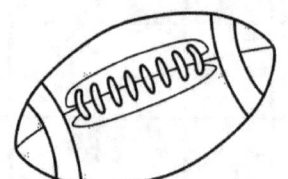 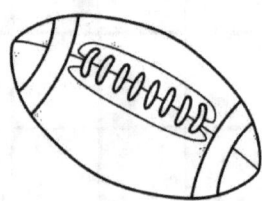

FOUR PROBLEMS IN A ROW: MULTIPLY BY 9
YOU NEED: CRAYONS OR COUNTERS

DIRECTIONS: On your turn, solve a problem and color it or cover it. Each player uses a different color. The first player to get 4 in a row wins!

9 X 12	5 X 9	9 X 9	3 X 9	9 X 10	4 X 9
9 X 5	10 X 9	12 X 9	9 X 4	8 X 9	9 X 2
7 X 9	9 X 8	9 X 0	3 X 9	4 X 9	1 X 9
6 X 9	9 X 11	9 X 7	9 X 6	1 X 9	11 X 9
9 X 2	5 X 9	9 X 9	8 X 9	10 X 9	9 X 3
9 X 12	11 X 9	9 X 6	9 X 9	9 X 5	4 X 9

POPPIN' PRODUCTS X9'S

YOU NEED: 1 DIE 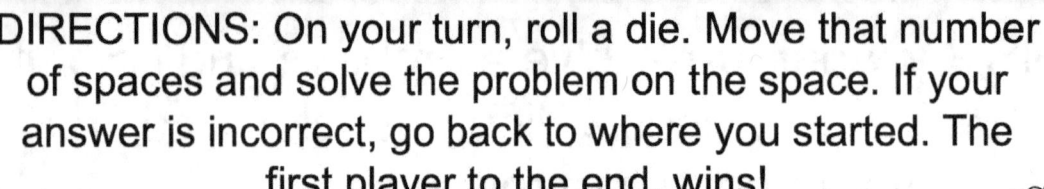 **COUNTERS**

DIRECTIONS: On your turn, roll a die. Move that number of spaces and solve the problem on the space. If your answer is incorrect, go back to where you started. The first player to the end, wins!

START

| 9X5 | 6X9 | 9X4 | 7X9 | 12X9 | 9X0 | 2X9 | 9X9 |

8X9

| 2X9 | 9X4 | 6X9 | 9X8 | 3X9 | 9X1 | | 9X3 |

| 9X10 | | | **END** | | 9X10 | | 10X9 |

| 1X9 | | 9X7 | | | 9X7 | | 1X9 |

| 9X2 | | 9X9 | 9X0 | 11X9 | 5X9 | | 11X9 |

| 9X11 | | | | | | | 9X5 |

| 8X9 | 9X7 | 9X12 | 0X9 | 9X6 | 9X9 | 9X3 | 4X9 |

© Laura Putman, Bright Minds Engaged, 2024-present, All rights reserved.

TRAP THE ICE CREAM: MULTIPLY BY 10
YOU NEED: COUNTERS

DIRECTIONS: On your turn, solve a problem. Put a counter on the problem. The last player who puts a counter around a cone traps it! Put a counter on that cone. See who can trap the most cones.

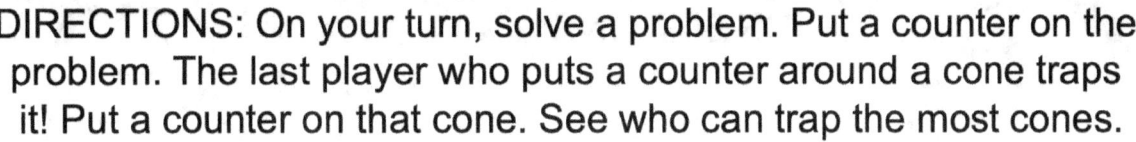

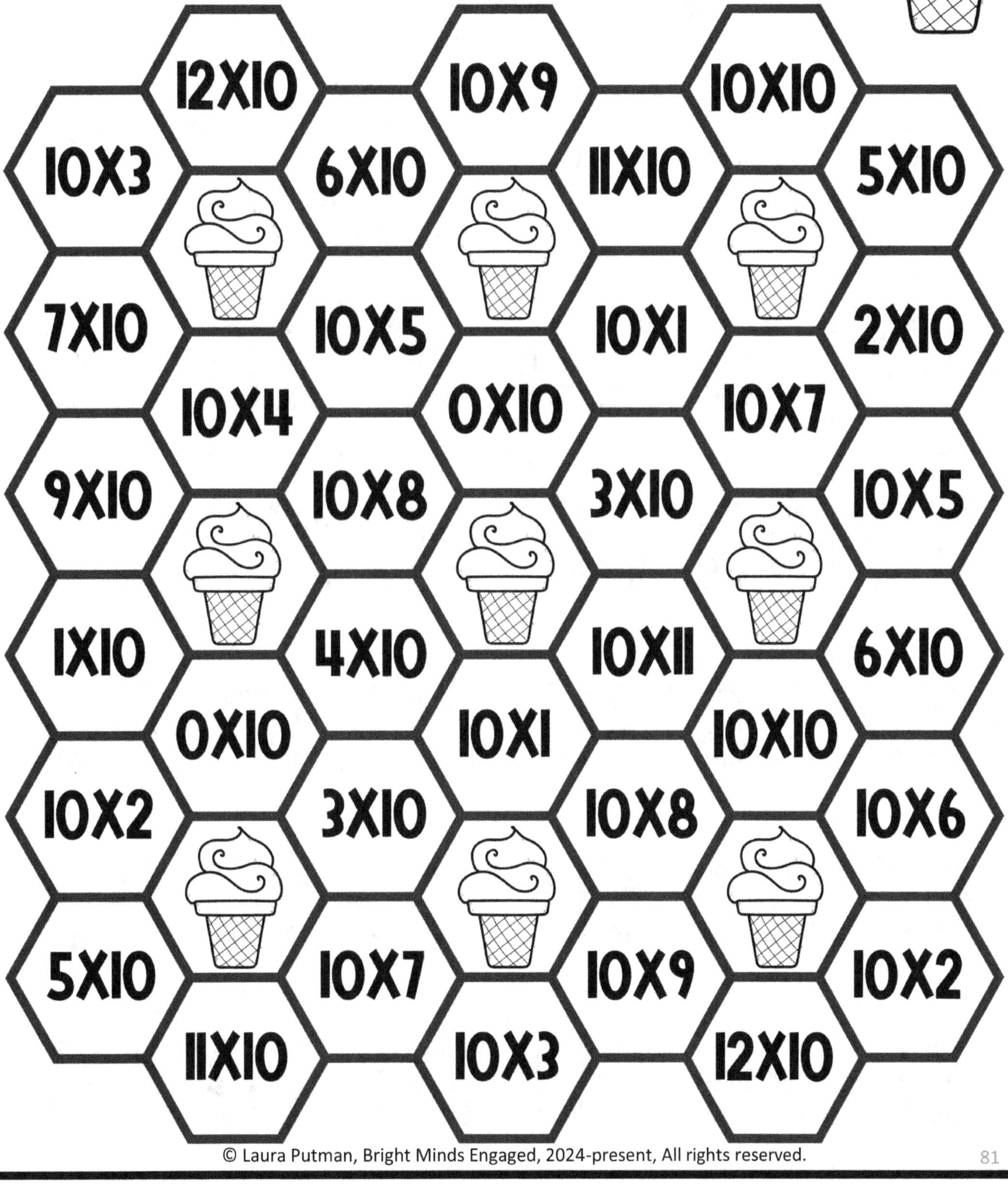

	12X10		10X9		10X10	
10X3		6X10		11X10		5X10
7X10		10X5		10X1		2X10
	10X4		0X10		10X7	
9X10		10X8		3X10		10X5
1X10		4X10		10X12		6X10
	0X10		10X1		10X10	
10X2		3X10		10X8		10X6
5X10		10X7		10X9		10X2
	11X10		10X3		12X10	

© Laura Putman, Bright Minds Engaged, 2024-present, All rights reserved.

SPIN A PROBLEM: MULTIPLY BY 10
YOU NEED: PAPERCLIP PENCIL CRAYONS

DIRECTIONS: Use a paperclip and pencil to make a spinner. On your turn, spin the paperclip. Multiply that number by 10 and color the product in the table. Each player uses a different color. If the product is not open, your turn is over. See who can solve the most problems!

10	70	120	40	20	80
40	90	60	50	100	30
110	20	30	80	10	40
60	120	50	100	90	110
50	30	90	20	70	60

PIG IN A PEN: MULTIPLY BY 10
YOU NEED: PAPERCLIP 📎 PENCIL ✏️

DIRECTIONS: On their turn, each player spins a number. Multiply the number by 10 and say the answer. If the answer is correct, draw a line to connect 2 dots. When a player completes a box, they write their initial in the box. At the end of the game, boxes are worth 1 point, and boxes with a pig in them are worth 5 points!

ARRAYS? HOORAY! MULTIPLY BY 10
YOU NEED: CRAYONS 2 DICE

DIRECTIONS: On their turn, each player rolls both dice. Add the dice together and multiply by 10. Draw an array for the problem. Each player uses a different color to make their arrays. Write the multiplication problem inside the array. If your array won't fit, your turn is over. When no more arrays can be made, the game is over. Whoever makes the most arrays, wins!

MULTIPLES OF 10 HUNT
YOU NEED: CRAYONS

Help the boy find his dog! Count by 10's to follow a path. If you get to 120, start at 10 again. Color or dab the spaces until you get to the bottom.

			START →	10	20	21	81	21	38	56	33
				18	14	30	20	84	81	92	99
				36	27	40	32	70	9	90	108
				60	50	56	72	74	110	52	9
24	48	80	80	70	40	54	60	40	62	11	19
63	56	90	44	109	80	72	35	54	49	29	27
64	110	100	9	108	99	10	60	112	45	36	29
120	23	18	7	50	24	90	30	54	80	72	38
10	20	30	40	60	36	81	72				
8	50	76	3	70	81	70	18				
16	102	54	80	64	90	3	94				
78	40	48	35	90	100	110	120	→ END			

FOUR PROBLEMS IN A ROW: MULTIPLY BY 10
YOU NEED: CRAYONS OR COUNTERS

DIRECTIONS: On your turn, solve a problem and color it or cover it. Each player uses a different color. The first player to get 4 in a row wins!

10 X 12	5 X 10	10 X 9	3 X 10	10 X 10	4 X 10
10 X 5	10 X 10	12 X 10	10 X 4	8 X 10	10 X 2
7 X 10	10 X 8	10 X 0	3 X 10	4 X 10	1 X 10
6 X 10	10 X 11	10 X 7	10 X 6	1 X 10	11 X 10
10 X 2	5 X 10	9 X 10	8 X 10	10 X 10	10 X 3
10 X 12	11 X 10	10 X 6	10 X 9	10 X 5	4 X 10

MATH BATTLE: X10'S

YOU NEED: 1 DIE **COUNTERS**

DIRECTIONS: On your turn, roll a die. Move that number of spaces and solve the problem on the space. If your answer is incorrect, go back to where you started. The first player to the end, wins!

START

10X3		10X0	8X10	10X7		10X9	6X10
7X10		9X10		10X10		3X10	**END** 
10X11		10X5		10X11		10X10	
4X10		10X2		2X10		4X10	
10X6		10X8		10X0		10X1	
10X10		12X10		1X10		7X10	
10X12		10X9		10X6		10X12	
1X10	5X10	4X10		8X10	3X10	10X2	

© Laura Putman, Bright Minds Engaged, 2024-present, All rights reserved.

ROLL & SOLVE: MULTIPLY BY 10 & 11

YOU NEED: 1 DIE 🎲 CRAYONS

DIRECTIONS: Assign one player even numbers on the die and the other player odd numbers. Take turns rolling. If a player rolls one of their numbers on the die, they solve the next problem under that die and color the space. If they do not roll one of their numbers, their turn is over. See who can fill their columns first!

⚀	⚁	⚂	⚃	⚄	⚅
8X11	10X5	7X10	2X10	10X8	11X3
10X12	11X11	10X11	8X11	11X9	0X10
5X10	10X8	4X10	3X10	12X11	7X11
1X11	1X10	11X0	4X11	10X7	11X4
10X4	9X10	5X11	10X10	11X2	10X6
8X10	12X11	10X6	11X3	6X11	10X9

© Laura Putman, Bright Minds Engaged, 2024-present, All rights reserved.

TIC-TAC-TOE: MULTIPLY BY 10 & 11

DIRECTIONS: Play a game of tic-tac-toe! Before you mark a space as yours, you must solve the problem in that space.

0X10	12X11	9X11
10X7	11X9	11X9
9X10	8X11	10X5

11X6	7X10	2X11
0X11	10X8	11X3
10X9	1X11	10X4

11X9	5X10	11X10
10X1	11X8	10X2
3X10	11X7	10X12

4X11	10X3	12X11
5X11	10X12	11X11
7X10	1X11	8X11

10X7	0X10	11X5
6X10	10X12	4X10
11X2	10X12	11X9

6X10	10X3	11X4
11X11	10X10	10X0
2X11	10X6	11X7

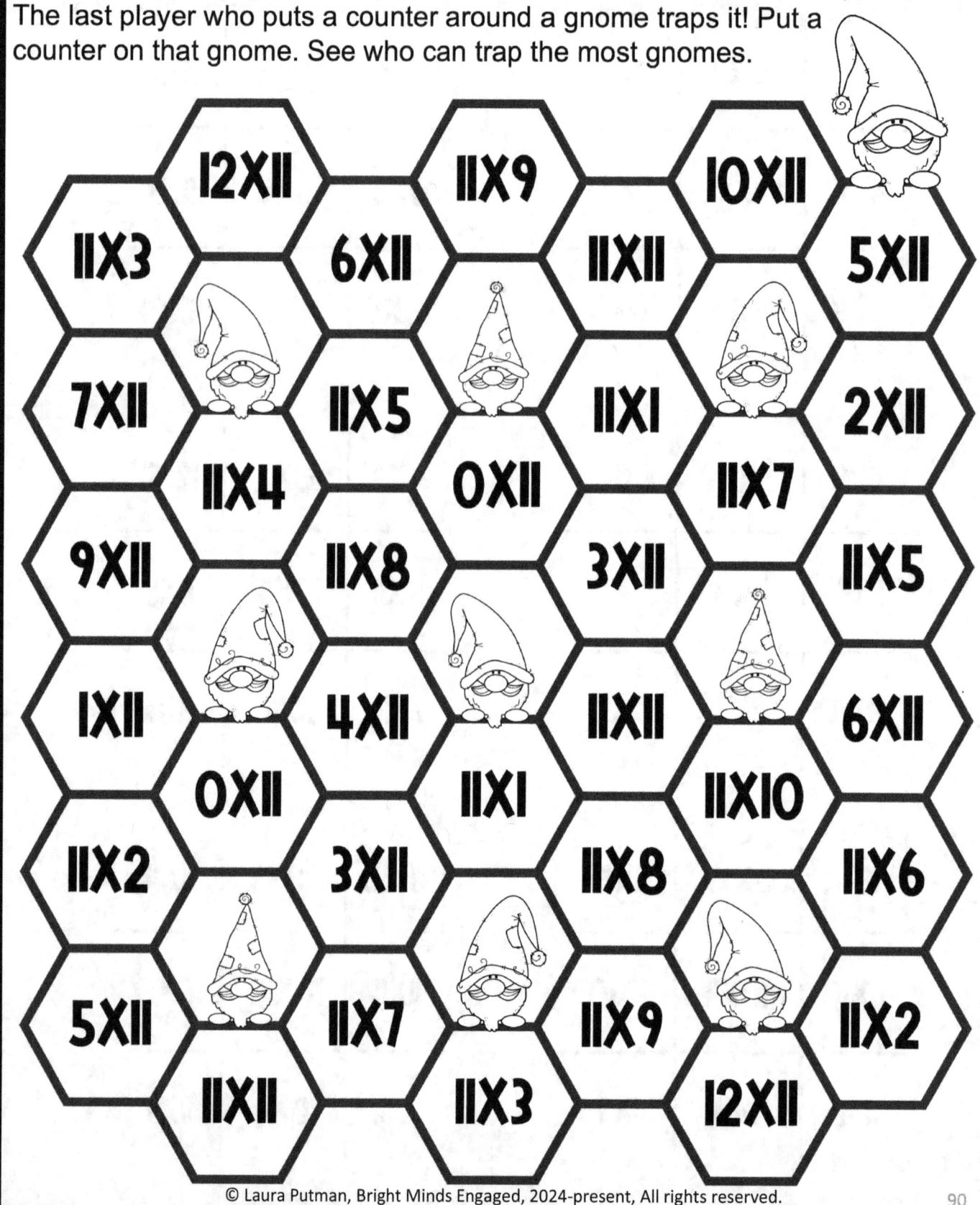

SPIN A PROBLEM: MULTIPLY BY 11
YOU NEED: PAPERCLIP 📎 PENCIL ✏️ CRAYONS

DIRECTIONS: Use a paperclip and pencil to make a spinner. On your turn, spin the paperclip. Multiply that number by 11 and color the product in the table. Each player uses a different color. If the product is not open, your turn is over. See who can solve the most problems!

11	77	132	44	22	88
44	99	66	55	110	33
121	22	33	88	11	44
66	132	55	110	99	121
55	33	99	22	77	66

© Laura Putman, Bright Minds Engaged, 2024-present, All rights reserved.

PIG IN A PEN: MULTIPLY BY 11
YOU NEED: PAPERCLIP 📎 PENCIL ✏️

DIRECTIONS: On their turn, each player spins a number. Multiply the number by 11 and say the answer. If the answer is correct, draw a line to connect 2 dots. When a player completes a box, they write their initial in the box. At the end of the game, boxes are worth 1 point, and boxes with a pig in them are worth 5 points!

MULTIPLES OF 11 HUNT
YOU NEED: CRAYONS

Help the honeybee get to its hive! Count by 11's to follow a path. If you get to 132, start at 11 again. Color or dab the spaces until you get to the bottom.

			START →	11	16	12	81	21	38	56	60
				13	22	33	40	84	66	92	99
				36	27	23	44	55	77	90	108
				11	45	10	41	88	36	52	9
24	48	80	66	27	121	54	99	42	101	18	19
63	55	40	44	132	80	110	112	54	49	29	27
64	72	44	9	108	11	19	96	90	45	110	29
21	33	18	77	24	27	22	33	54	80	72	38
96	22	34	40	48	36	44	72				
8	30	44	76	72	55	74	77				
16	52	54	63	64	90	66	88				
88	40	48	35	71	78	110	99	→ END			

© Laura Putman, Bright Minds Engaged, 2024-present, All rights reserved.

FOUR PROBLEMS IN A ROW: MULTIPLY BY 11

YOU NEED: CRAYONS OR COUNTERS

DIRECTIONS: On your turn, solve a problem and color it or cover it. Each player uses a different color. The first player to get 4 in a row wins!

11 X 12	5 X 11	11 X 9	3 X 11	10 X 11	4 X 11
2 X 11	0 X 11	11 X 9	6 X 11	11 X 1	11 X 7
7 X 11	11 X 8	11 X 0	3 X 11	4 X 11	1 X 11
6 X 11	11 X 11	11 X 7	11 X 6	1 X 11	11 X 11
11 X 2	5 X 11	9 X 11	8 X 11	11 X 10	11 X 3
11 X 12	11 X 11	11 X 6	11 X 9	11 X 5	4 X 11

© Laura Putman, Bright Minds Engaged, 2024-present, All rights reserved.

MULTIPLICATION MYSTERY XII'S

YOU NEED: 1 DIE COUNTERS

DIRECTIONS: On your turn, roll a die. Move that number of spaces and solve the problem on the space. If your answer is incorrect, go back to where you started. The first player to the end, wins!

START

11X3		11X0	8X11	7X11		11X9	11X6
7X11		9X11		10X11		3X11	END
11X11		11X5		11X11		11X10	
4X11		2X11		2X11		4X11	
11X6		11X8		11X0		11X1	
10X11		12X11		1X11		7X11	
11X12		11X9		11X6		11X11	
1X11	5X11	11X4		8X11	3X11	11X2	

ROLL & SOLVE: MULTIPLY BY 12
YOU NEED: 1 DIE • CRAYONS

DIRECTIONS: Assign one player even numbers on the die and the other player odd numbers. Take turns rolling. If a player rolls one of their numbers on the die, they solve the next problem under that die and color the space. If they do not roll one of their numbers, their turn is over. See who can fill their columns first!

⚀	⚁	⚂	⚃	⚄	⚅
12X1	5X12	12X7	12X10	8X12	2X12
12X12	12X11	10X12	12X8	3X12	0X12
5X12	12X8	12X4	12X3	12X12	12X7
12X2	1X12	12X9	4X12	7X12	12X9
4X12	12X2	12X5	11X12	2X12	12X6
12X0	12X12	6X12	12X3	12X6	11X12

TRAP THE MICE: MULTIPLY BY 12
YOU NEED: COUNTERS

DIRECTIONS: On your turn, solve a problem. Put a counter on the problem. The last player who puts a counter around a mouse traps it! Put a counter on that mouse. See who can trap the most mice.

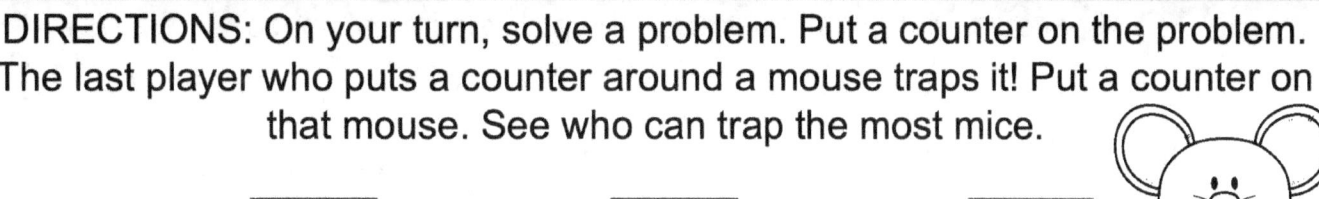

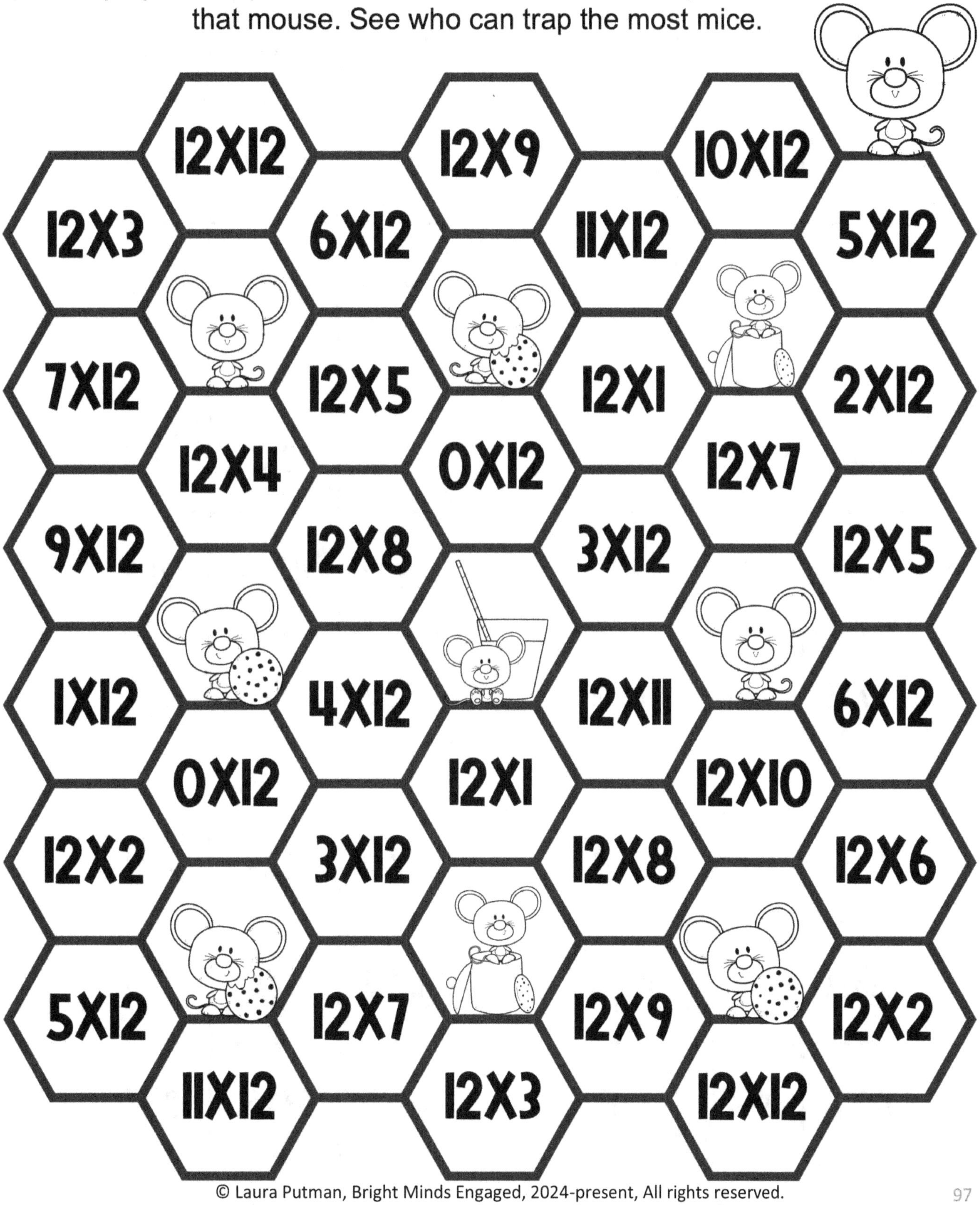

- 12X12
- 12X9
- 10X12
- 12X3
- 6X12
- 11X12
- 5X12
- 7X12
- 12X5
- 12X1
- 2X12
- 12X4
- 0X12
- 12X7
- 9X12
- 12X8
- 3X12
- 12X5
- 1X12
- 4X12
- 12X11
- 6X12
- 0X12
- 12X1
- 12X10
- 12X2
- 3X12
- 12X8
- 12X6
- 5X12
- 12X7
- 12X9
- 12X2
- 11X12
- 12X3
- 12X12

© Laura Putman, Bright Minds Engaged, 2024-present, All rights reserved.

TIC-TAC-TOE: MULTIPLY BY 12

DIRECTIONS: Play a game of tic-tac-toe! Before you mark a space as yours, you must solve the problem in that space.

0X12	12X12	12X1
12X7	12X9	11X12
12X10	8X12	12X5

12X6	7X12	2X12
0X12	12X8	12X3
9X12	1X12	12X4

12X9	5X12	12X10
12X1	12X8	12X2
3X12	12X7	12X12

4X12	12X3	12X12
5X12	12X12	12X11
7X12	1X12	8X12

12X7	0X12	12X5
6X12	10X12	4X12
12X2	12X11	3X12

6X12	12X3	12X4
11X12	10X12	12X0
12X9	12X6	12X7

SPIN A PROBLEM: MULTIPLY BY 12
YOU NEED: PAPERCLIP ✏️ PENCIL ✏️ CRAYONS

DIRECTIONS: Use a paperclip and pencil to make a spinner. On your turn, spin the paperclip. Multiply that number by 12 and color the product in the table. Each player uses a different color. If the product is not open, your turn is over. See who can solve the most problems!

12	84	144	48	24	96
48	108	72	60	120	36
132	24	36	96	12	48
72	144	60	120	108	132
60	36	108	24	84	72

© Laura Putman, Bright Minds Engaged, 2024-present, All rights reserved.

PIG IN A PEN: MULTIPLY BY 12
YOU NEED: PAPERCLIP 🖇 PENCIL ✏️

DIRECTIONS: On their turn, each player spins a number. Multiply the number by 12 and say the answer. If the answer is correct, draw a line to connect 2 dots. When a player completes a box, they write their initial in the box. At the end of the game, boxes are worth 1 point, and boxes with a pig in them are worth 5 points!

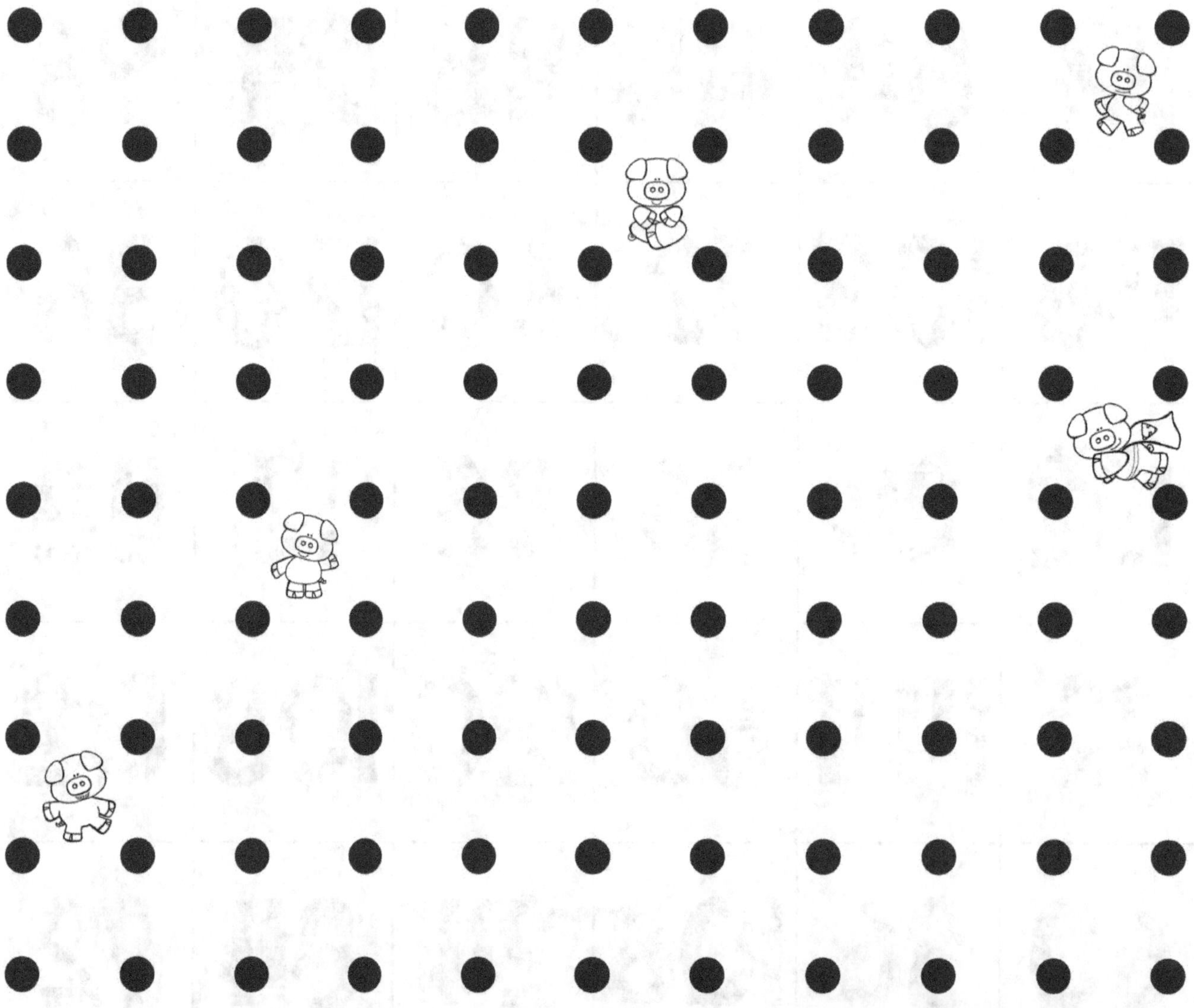

MULTIPLES OF 12 HUNT
YOU NEED: CRAYONS

Help the chef deliver the pizza! Count by 12's to follow a path. If you get to 144, start at 12 again. Color or dab the spaces until you get to the bottom.

			START →	12	24	30	48	21	38	56	42
				20	14	36	60	84	81	92	99
				36	27	23	32	72	84	90	108
				6	45	56	63	100	34	96	110
22	48	80	60	48	44	54	64	112	120	108	19
63	56	40	72	109	36	72	35	144	132	29	27
64	72	84	9	108	99	24	12	56	45	36	29
21	96	18	7	24	77	90	63	54	80	72	38
108	27	34	30	48	36	81	72				
8	120	45	24	72	81	45	18				
132	144	12	63	36	92	60	74				
78	40	48	35	71	48	112	72	END →			

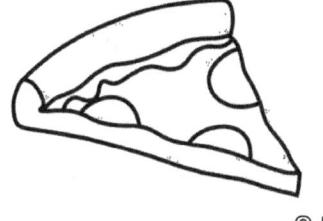

FOUR PROBLEMS IN A ROW: MULTIPLY BY 12
YOU NEED: CRAYONS OR COUNTERS

DIRECTIONS: On your turn, solve a problem and color it or cover it. Each player uses a different color. The first player to get 4 in a row wins!

12 X 12	5 X 12	12 X 9	3 X 12	10 X 12	4 X 12
2 X 12	0 X 12	12 X 9	6 X 12	12 X 1	12 X 7
12 X 5	12 X 10	12 X 12	12 X 4	8 X 12	12 X 2
7 X 12	12 X 8	12 X 0	3 X 12	4 X 12	1 X 12
6 X 12	11 X 12	12 X 7	12 X 6	1 X 12	12 X 11
12 X 2	5 X 12	9 X 12	8 X 12	12 X 10	12 X 3

FOOTBALL FACTS X12'S

YOU NEED: 1 DIE COUNTERS

DIRECTIONS: On your turn, roll a die. Move that number of spaces and solve the problem on the space. If your answer is incorrect, go back to where you started. The first player to the end, wins!

START

| 5X12 | 6X12 | 12X4 | 7X12 | 12X12 | 12X0 | 2X12 | 12X9 |

| | | | | | | | 8X12 |

| 2X12 | 12X4 | 6X12 | 12X8 | 3X12 | 12X1 | 22 | 12X3 |

| 12X10 | 44 | **END** | | 12X10 | | 10X12 |

| 1X12 | | 7X12 | | 4X12 | | 1X12 |

| 12X2 |  | 12X9 | 12X0 | 11X12 | 5X12 | 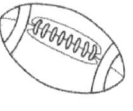 | 11X12 |

| 12X11 | | | | | | | 5X12 |

| 8X12 | 12X7 | 12X12 | 0X12 | 12X6 | 12X9 | 12X3 | 4X12 |

© Laura Putman, Bright Minds Engaged, 2024-present, All rights reserved.

ROLL & SOLVE: MIXED FACTS
YOU NEED: 1 DIE 🎲 CRAYONS

DIRECTIONS: Assign one player even numbers on the die and the other player odd numbers. Take turns rolling. If a player rolls one of their numbers on the die, they solve the next problem under that die and color the space. If they do not roll one of their numbers, their turn is over. See who can fill their columns first!

⚀	⚁	⚂	⚃	⚄	⚅
4X4	0X5	7X3	5X5	8X9	8X3
12X6	7X7	6X10	2X8	4X5	0X8
5X6	12X3	3X4	3X3	12X10	6X3
7X5	1X4	0X11	4X6	2X7	5X9
9X9	9X10	8X8	8X11	2X2	2X4
0X10	8X5	12X1	9X3	7X6	11X10

TRAP THE MONSTERS: MIXED FACTS
YOU NEED: COUNTERS

DIRECTIONS: On your turn, solve a problem. Put a counter on the problem. The last player who puts a counter around a monster traps it! Put a counter on that monster. See who can trap the most monsters.

- 12X1, 9X9, 4X10
- 3X0, 6X0, 11X12, 5X3
- 6X3, 12X10, 5X2, 11X12
- 10X10, 10X9, 7X8
- 9X6, 4X8, 1X3, 5X7
- 2X2, 7X4, 5X11, 6X8
- 0X0, 7X7, 4X4
- 5X9, 2X4, 8X8, 4X6
- 6X5, 6X7, 8X9, 2X3
- 1X0, 3X4, 5X5

MIXED FACTS TIC-TAC-TOE

DIRECTIONS: Play a game of tic-tac-toe! Before you mark a space as yours, you must solve the problem in that space.

6X3	8X8	12X4		6X6	7X10	11X2
2X5	7X9	11X11		0X12	5X6	3X8
1X10	9X8	5X4		9X10	7X2	3X4

8X6	5X5	12X11		9X2	3X3	0X0
4X9	5X8	6X10		5X12	9X3	4X11
2X3	4X4	1X2		7X7	6X2	8X1

7X5	8X2	3X5		8X2	1X3	4X2
6X6	10X10	9X5		7X4	5X10	12X9
4X8	12X12	3X7		11X5	6X4	9X9

SPIN A PROBLEM: MIXED FACTS
YOU NEED: PAPERCLIP 📎 PENCIL ✏️ CRAYONS

DIRECTIONS: Use a paperclip and pencil to make a spinner. On your turn, spin the paperclip. Multiply that number by a number of your choice in the grid and color it. Each player uses a different color. See who can solve the most problems!

1	7	12	4	2	8
4	9	6	5	10	3
11	2	0	8	1	4
6	12	5	10	9	11
0	3	9	2	7	6

PIG IN A PEN: MIXED FACTS
YOU NEED: PAPERCLIP / PENCIL / 2 DICE

DIRECTIONS: On their turn, each player rolls both dice and spins a number. Multiply the numbers. If the answer is correct, draw a line to connect 2 dots. When a player completes a box, they write their initial in the box. At the end of the game, boxes are worth 1 point, and boxes with a pig in them are worth 5 points!

ARRAYS? HOORAY! MIXED FACTS

YOU NEED: CRAYONS 2 DICE

DIRECTIONS: On their turn, each player rolls both dice twice. Make an array using the number you rolled. Each player uses a different color to make their arrays. Write the multiplication problem inside the array. If your array won't fit, your turn is over. When no more arrays can be made, the game is over. Whoever makes the most arrays, wins!

FOUR PROBLEMS IN A ROW: MIXED FACTS
YOU NEED: CRAYONS OR COUNTERS

DIRECTIONS: On your turn, solve a problem and color it or cover it. Each player uses a different color. The first player to get 4 in a row wins!

2 X 2	0 X 8	9 X 9	6 X 4	2 X 1	5 X 7
5 X 3	10 X 10	12 X 3	4 X 9	8 X 8	2 X 8
7 X 7	4 X 8	0 X 0	9 X 3	5 X 4	1 X 3
6 X 5	12 X 11	3 X 7	6 X 6	1 X 1	11 X 2
4 X 4	5 X 5	7 X 6	5 X 9	10 X 8	3 X 2
10 X 12	3 X 4	6 X 8	9 X 2	2 X 5	4 X 1

POPPIN' PRODUCTS: MIXED FACTS
YOU NEED: 1 DIE 🎲 COUNTERS ⚫⚫

DIRECTIONS: On your turn, roll a die. Move that number of spaces and solve the problem on the space. If your answer is incorrect, go back to where you started. The first player to the end, wins!

START

| 5X3 | 6X8 | 4X4 | 7X4 | 12X10 | 0X0 | 2X3 | 8X9 |

7X7

| 10X10 | 2X4 | 6X3 | 7X8 | 2X2 | 12X1 | | 4X8 |

| 8X8 | | | | | 4X10 | | 10X9 |

END

| 6X0 | | 6X7 | | | 5X7 | | 1X3 |

| 5X2 | | 3X4 | 3X0 | 11X12 | 4X3 | | 1X0 |

| 11X11 | | | | | | | 5X11 |

| 8X8 | 3X7 | 9X9 | 6X5 | 9X6 | 5X9 | 5X5 | 4X6 |

© Laura Putman, Bright Minds Engaged, 2024-present, All rights reserved.

MULTIPLICATION FACTS
Parent Strategy Guide

Get the most out of this workbook!

✓ Tips for Success
✓ Strategy Posters
✓ Progress Chart

Scan here to get it!

The guide to helping your child master multiplication facts

By Laura Putman, M.Ed

© Laura Putman, Bright Minds Engaged, 2026-present, All rights reserved.

Does your child need extra practice in multiplication facts?

If your child is benefiting from our Long Division Workbook, you can get our **Multiplication Facts Workbook** designed to engage kids who struggle with learning multiplication facts.

Multiplication Facts Workbook

This workbook is perfect for:
- ✔ Kids who struggle with math facts
- ✔ Ages 8-12
- ✔ Reinforcing multiplication facts
- ✔ Kids who need math motivation
- ✔ Busy parents
- ✔ Math practice at home

It includes:
- ✔ A variety of pages to engage kids
- ✔ Repeated practice to build facts
- ✔ An easy-to-use answer key
- ✔ Kid-friendly graphics
- ✔ Enough practice for real mastery

⭐⭐⭐⭐⭐

Used by teachers, tutors, parents, and homeschool families to build confidence and close learning gaps.

Scan to buy on Amazon.

©2026 Laura Putman, Bright Minds Engaged